givi
nat
a h

Snakes

Jules Howard

BLOOMSBURY WILDLIFE
LONDON · OXFORD · NEW YORK · NEW DELHI · SYDNEY

BLOOMSBURY WILDLIFE
Bloomsbury Publishing Plc
50 Bedford Square, London, WC1B 3DP, UK

BLOOMSBURY, BLOOMSBURY WILDLIFE and the Diana logo are trademarks of
Bloomsbury Publishing Plc

First published in United Kingdom 2020

A catalogue record for this book is available from the British Library

Library of Congress Cataloguing-in-Publication data has been applied for

ISBN: PB: 978-1-4729-7169-2; ePub: 978-1-4729-7168-5; ePDF: 978-1-4729-7167-8

2 4 6 8 10 9 7 5 3 1

Design by Rod Teasdale
Printed and bound in India by Replika Press Pvt. Ltd.

MIX
Paper from
responsible sources
FSC® C016779

To find out more about our authors and books visit www.bloomsbury.com
and sign up for our newsletters

giving
nature
a home

Published under licence from RSPB Sales Limited to raise awareness of the RSPB (charity
registration in England and Wales no 207076 and Scotland no SC037654).

For all licensed products sold by Bloomsbury Publishing Limited, Bloomsbury Publishing Limited
will donate a minimum of 2% from all sales to RSPB Sales Ltd, which gives all its
distributable profits through Gift Aid to the RSPB.

Contents

The Wonder of Snakes

No group of animals has taken to life without legs quite like snakes. For more than 150 million years these dynamic predatory reptiles have adapted to, and exploited, a variety of habitats. Today, more than 3,000 snake species live on Earth, and each has its own distinctive markings and patterns, its own hunting styles, its own unique courtship behaviours and its own seasonal habits. This introduction explores what makes snakes such an enduring and dramatic evolutionary success story.

Perhaps no other animal in Britain elicits such a variety of responses than snakes. To some, they are organisms that instil a deep wonder in the natural world. To others, they are symbols of a dwindling wild landscape that is slowly fading from view. To many, sadly, they are objects to loathe, avoid or hate. I should know, for I used to manage a phone helpline that offered advice to lovers and loathers of snakes in Britain. From my work on this service, it became apparent to me that there is a real need for more sources of clear advice about British snakes – including identification and the increasing need for their conservation, which I hope this book will deliver.

But I'm keen to offer you, the reader, a bit more. My aim is that you finish this book as an even more pronounced lover of snakes – someone with an understanding about what these animals mean for our cultural history and, if you aren't already, someone with a spring in your step to take part in their conservation. No other creature has been more maligned and misunderstood than British snakes. I hope that, together, we can right that wrong.

Above: Even though snakes have been a feature of British ecosystems for centuries, their secretive nature means that relatively few people in Britain have ever seen one.

Opposite: The Adder is one of Britain's most renowned snakes. It features in art, poetry and folklore. Predictably, not all representations of this venomous snake are positive.

Snake origins

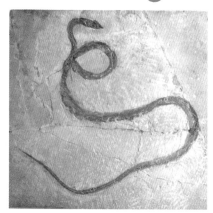

Above: Because snake bones are more delicate than those of many other terrestrial vertebrates, their fossils are comparatively harder to find. That makes studying snake evolution more taxing than for many vertebrates.

Below: Snakes aren't the only reptiles that have lost their legs over time. Amphisbaenians (left) and Slow Worms (legless lizards) (right) have seen similar burrowing adaptations.

The evolutionary history of snakes is a suitably convoluted affair. Each year, new fossil discoveries push the origin of snakes further and further back in the story of life on Earth. Traditionally, snakes were considered to have lived alongside the last of the dinosaurs in the Cretaceous Period, 100 million years ago. New fossil evidence then pushed their origins further back in time, perhaps to the Jurassic Period, 150 million years ago. More recently, in 2015, fragmentary fossils from museum specimens (including some collected from southern England) were re-examined, suggesting that snakes were not only around 167 million years ago but had already diversified and were living in a host of habitats, much as they are today. This suggests that the origins of snakes may go back further still – some scientists believe they may have evolved among the early dinosaurs of the Triassic Period, more than 200 million years ago.

With their scaly skin and peculiar anatomical arrangements, snakes are clearly reptiles, specifically belonging to a group (technically called an order) of reptiles known as squamates – scaly-skinned reptiles. But from which early squamate did snakes descend? The first snakes are thought to have evolved from a group of early lizards whose legs gradually became shorter and shorter over successive generations – probably as an

adaptation to a life of burrowing. Many animals that burrow tend to evolve a streamlined shape to allow them to move more easily through soil or sand, and this often means they lose their legs. This adaptation is seen in caecilians (a group of amphibians), in amphisbaenians (a group of worm-like lizards) and in some modern-day lizards, such as Slow Worms (*Anguis fragilis*). Leglessness, it seems, is one of the more common animal adaptations when it comes to a life of burrowing.

However, the evolutionary story of snakes may not be this simple. Some scientists have argued that snakes may have had their origins in the oceans, starting out as land-living reptiles, evolving into a streamlined, legless swimming form, and then further evolving back into a land-living world-beater. As enticing as this idea seems, the transitional fossils of early snakes (complete with tiny legs) indicate that their legs were probably a terrestrial adaptation for digging.

Some snake groups, like boas and pythons, have retained their vestigial hind limbs as small claw-like protrusions called 'spurs', which perform a role in mating. Within these spurs are the remains of the ilium and the femur bones of their four-legged ancestors.

All of this means that, in the same way that scientists now consider birds to be the descendants of dinosaurs, we can consider snakes to form a highly specialised early branch of the lizard family tree. And like birds, they have had a dramatic impact on life on Earth.

Above: Some scientists argue that snakes may have evolved at sea and later evolved back into terrestrial animals. However, recent evidence suggests this is unlikely.

Below: The 'pelvic spurs' of boas and pythons are vestigial limbs. Many species use them as a tickling aid during copulation.

The diversity of snakes

Above: Pythons are celebrated and well-known predators. They use their powerful jaws and muscular coils to capture and restrain unsuspecting prey.

Below: In comparison to the python above, the Little Wart Snake is one of many over-looked snakes. Its tough, baggy skin helps it to constrict prey underwater.

Experts have described more than 3,000 snake species, although there are likely to be many more out there yet to be discovered. In all, the group contains 20 taxonomic families, some of which are well known to anyone who watches and enjoys nature documentaries. The celebrities of the snake world include the pythons (*Pythonidae*), the boas (*Boidae*) and the vipers (*Viperidae*). But there are also many weird and wonderful families that receive less of the limelight. These include the rather primitive-looking wart snakes (*Acrochordidae*), the highly iridescent sunbeam snakes (*Xenopeltis*) and the diminutive and almost dainty snail-eating snakes (*Pareidae*) of Southeast Asia.

Perhaps the most diverse thing about snakes is their size. Among the smallest species is the Barbados Thread Snake (*Tetracheilostoma carlae*), found on the Caribbean island of Barbados and measuring just 10.4cm (4in), and the Brahminy Blind Snake (*Indotyphlops braminus*), said

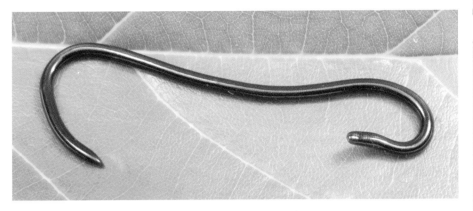

to measure a mere 10.2cm (4in). The latter is also one of the most wide-roving snakes, thanks to its habit of finding its way into plant pots being readied for international distribution. Once native to Africa, the species is now found in the Americas, Australia and Oceania. For such a widespread species, it is interesting that not a single male Brahminy Blind Snake has ever been found. This makes it one of a number of parthenogenetic snake species, in which the females produce clones of themselves in each batch of eggs rather than seeking out mates. Snakes are nothing if not inventive, evolutionarily speaking.

Above: The Brahminy Blind Snake regularly travels the world in flowerpots. To tell it from a worm, look for its tiny scales.

Below: Like all boas, the Tartar Sand Boa (*Eryx tataricus*) possesses a more rigid lower jaw than other snakes, which is a feature considered primitive among snakes.

Above: The Reticulated Python – the modern age's longest snake according to the Guinness Book of World Records.

Below: Though pythons (left) and anacondas (right) can get very large, many individuals in a population are sub-adults. Therefore not every snake encounter is with an enormous individual.

The most bulky and lengthy snakes alive today are mostly from two families, the pythons and the boas. The award for longest snake surely goes to the Southeast Asian Reticulated Python (*Python reticulatus*), which regularly reaches lengths of 5–7m (16.5–23ft), although many wild specimens that are even longer than this have been reported. The Green Anaconda (*Eunectes murinus*) measures slightly less in total length (nearer 5m, or 16.5ft), but it more than makes up for this in terms of its bulk. A Reticulated Python the same length as an adult Green Anaconda may be as little as half its weight.

Despite their impressive size, pythons and boas are dwarfed by the snakes of prehistory. For example, fossils of *Titanoboa* species provide evidence that some snakes adequately filled the role of apex predator in the 10 million years after the demise of the dinosaurs, at least in South America. *Titanoboa cerrejonensis* reached a formidable 12.8m (42ft) in length, a fact that any dinosaur-obsessed child will be eager to tell you.

Above left: Many snakes, such as this green pit viper, are highly mobile and easily capable of scaling the tallest of branches.

Above right: A uniquely colourful Ball Python (*Python regius*). As well as differing in size and stature, snake species differ widely in their colours and markings.

Left: An adult Green Anaconda may weigh as much as 70kg (11st).

Ecosystem exploiters

Above: Sea snakes are one of the world's most elegant hunters. Their thin paddle-like tail (visible here) allows them to glide almost effortlessly through water.

Below: Many snake species (including the so-called 'flying' snakes) have adapted to life in the tree tops where new food sources, such as birds and their chicks, can be exploited.

When we look at snakes, we see them as having a single body plan, moulded and adapted for a multitude of lifestyles.

Among the world's most successful snakes are those that make their home in water. Of the sea snakes (mostly comprising a group called the Hydrophiinae, and found in coastal areas of the Indian and Pacific oceans), there are 69 known species, each with a paddled tail and an eel-like body that allows it to slither in and out of cracks between rocks and corals. Included in their ranks are some of the world's most venomous species.

Although most snakes live on the ground, many species have come to explore other ecological niches. These include tree-living (arboreal) snakes such as tree boas and tree pythons, which coil around branches and wait patiently to ambush prey. Then there are the so-called 'flying' snakes, members of the South Asian genus *Chrysopelea*, which can expand their ribs to create a simple aerofoil or pseudo-concave wing. Some flying snakes have been observed gliding more than 100m (330ft) between trees when spooked by predators, outcompeting more famous gliders such as flying squirrels and dragon lizards.

Some snakes have returned to a life in the soil, such as the Sri Lankan Blyth's Earth Snake (*Rhinophis blythii*), with a pointed, armoured skull and degenerative eyes. Many other species also favour dark places, none more so than cave-dwelling populations of the Yellow-red Rat Snake (*Pseudelaphe flavirufa*), which is known to dangle from the roofs of some caves in its native

Above: The Emerald Tree Boa (*Corallus caninus*) is an attractive arboreal hunter known from the rainforests of South America.

Mexico and Central America to strike at passing bats. There are agile and athletic snakes, such as the Black Mamba (*Dendroaspis polylepis*) of sub-Saharan Africa, which can move after prey in short bursts of up to 10mph (16kph). And there are sit-and-wait specialists, such as the rattlesnakes, which use venom to incapacitate their prey. As is well known, venomous snakes can often use their venom as a means of defending themselves against threats, including humans. Some, like the group of snakes known as spitting cobras, can squirt their venom at would-be attackers with impressive accuracy.

Below: Only a single small population of Yellow-red Rat Snakes have taken up the curious habit of hunting bats. The behaviour might also occur in other, as yet undiscovered, caves across its Central American range.

Snakes really do get around, geographically. Although many species live in the tropics, some are specially adapted to more temperate climates. The Adder (*Vipera berus*), for example, has adapted to cope with near-freezing spring temperatures – hence its wide ranging presence throughout England, Wales and Scotland. In fact, the species is so well adapted to the cold that it can even occupy environments within the Arctic Circle in some parts of its range.

Hither and slither

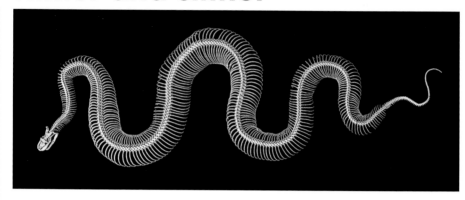

Above: The most notable feature of a snake's skeleton is the numerous vertebrae and the strength and flexibility of their attachments. Some species have 400 vertebrae in total.

Below: The scales on the underside of a snake's body are highly sensitive to touch – they help the snake feel for potential 'push-points' to maintain their forward motion.

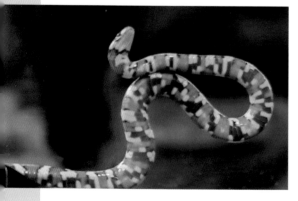

The success of snakes is, in great part, down to their unique musculoskeletal system. Where humans have 12 vertebrae in the middle (thoracic region) of the spine, snakes can have more than a hundred. These bones and their associated muscles and tendons allow for the charismatic S-shaped motions (classically described as serpentine movement) that for most snakes are the primary form of locomotion.

Scales are important in serpentine movement, as those on the underside of the snake's body are highly sensory. Each scale passes information to the brain about physical objects such as rocks, cracks or sturdy twigs that the snake can push off to maintain forward motion. In serpentine movement, the propulsive force comes from the sides of the snake's body – this is what makes for the classic slithering motion. The drawback to this movement mechanism is that, should a snake find itself on a polished and featureless plain, it will struggle to gain traction and its rate of travel will be dramatically limited. Thankfully, nature is anything but featureless. Even a tiny hillock or a pebble can provide a sturdy push point (more formally called a resistance site) from which forward propulsion can be attained. Interestingly, snakes use this method of locomotion most rhythmically when in water. Here, the water itself provides the medium from which the snake can push off, evident in its S-shaped wave of motion.

Other methods through which snakes can move include sidewinding or a so-called 'concertina' movement. In sidewinding, snakes use their head and tail to push downwards against the substrate, making their own resistance site from which they fling themselves forward in a sideways-like motion. This method of movement suits species that live in featureless arid environments, such as deserts. In concertina movements, the snake anchors a part of its body against the substrate, often gripping with its lowermost (ventral) scales, before stretching the body forward to seek new anchor points from which to grip and move off.

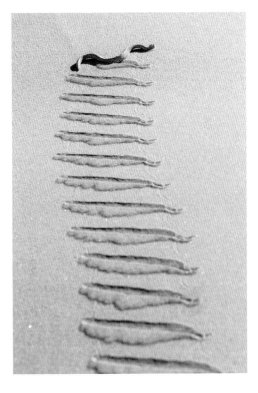

Other ways snakes move include the caterpillar locomotion employed by some larger species. Rather than winding their bodies from side to side, these snakes can undulate up and down with a series of tiny ripples that run in waves down the body, making them appear worm-like. To passers-by, this form of snake movement looks almost supernatural.

Above and below: Through its unusual method of locomotion, the Sidewinder Rattlesnake can maintain traction with the floor while moving across windblown desert sands. This results in a classic 'j-shaped' trackway.

What natural selection – the engine of evolution – takes away with one hand, it can give back with the other. This is especially the case with snakes, for which the loss of legs has provided for the evolution of an extraordinary skeleton and a sensory system par excellence. As well as propulsion, the well-muscled and incredibly coordinated movement of snakes allows for some other

Above: The Mexican Burrowing Snake uses its narrow, shovel-shaped head to burrow through soil. Like other burrowing vertebrates, its eyes are very small.

Below: The muscular body, stream-lined head and smooth scales of the Western Shovelnose Snake allow it to 'swim' through sand.

impressive feats, including digging, leaping and climbing.

When temperatures get too cold or too hot for comfort, some snakes simply choose to bury themselves in a soft substrate to seek shelter from the elements. They manage this seemingly impossible act by pressing their coils downwards and churning up the soil or sand, which ends up on top of the snake as it digs downwards. Some snakes, such as the Sidewinder Rattlesnake (*Crotalus cerastes*), employ this technique for ambush hunting – when buried in the sand, only their eyes and mouth remain visible. Other snakes are even more specialised diggers. The Mexican Burrowing Snake (*Loxocemus bicolor*) has a long bucket-like jaw that it uses to burrow through soil. Many burrowing species push through the soil using a concertina-like motion and some, such as the Western Shovelnose Snake (*Chionactis occipitalis*), which is found mainly in western California, actually manage to 'swim' through loose sand while hunting or keeping out of predators' reach.

Although not an attribute seen among British species, 'leaping' short distances is a trick employed by some snakes to surprise would-be attackers. This jumping motion is normally the result of a mock bite to scare away a predator – the snake lunges its jaws forward with such momentum that the head and the tail are momentarily launched into the air. These 'leaps' can be unerringly convincing on downward-sloping ground or atop boulders or branches. The most famous springing snakes are the jumping pit vipers, species of the *Atropoides* genus from Central America and Mexico. These species are said to leap more than a full body length at passing prey, although some scientists remain a little dubious about these claims.

Snakes that climb achieve this feat through a concertina movement. Like human climbers with crampons, they look out for sturdy holdfasts, which they grasp with their scales rather than with hands and feet. At other times, snakes move between tree branches using each branch and twig as a resistance point from which to push further up into the tree. In Britain, the Barred Grass Snake (*Natrix helvetica*) frequently uses this technique when scaling the upper branches of hedgerows in search of nestlings, for instance.

Above: Even an apparently featureless tree is covered in hidden bumps and gnarled edges that snakes can employ to push off and move upwards.

Left: The leaping ability of the so-called 'jumping pit vipers' is often over-hyped, but springing toward prey clearly brings adaptive benefits by way of food.

Snake senses

Compared to other predators, snakes are almost unmatched in nature in relation to their sensory endowments, and their senses of smell, touch and vision are particularly impressive. Yet there are mysteries around some of their sensory adaptations that scientists are only now beginning to unpick. Some of these impressive adaptations are explored below.

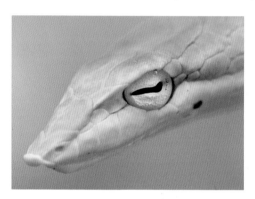

Above: Horizontal pupils are rare among snakes. They allow greater depth of field in the horizontal plane but still allow in enough light for the snake to hunt at night.

Below: Famously, the snake's forked tongue is double-pronged. It can pick up and process smells in stereo, something we are not able to do.

Sight

You can learn a lot about a snake by looking it straight in the eyes – or rather, by looking it in the eye and studying its pupils. Snake pupils come in three basic shapes: round, vertical and horizontal. Round pupils (like our own) are good at seeing objects in the light of day, and are especially common in snakes that regularly move through water. Vertical pupils are more common in species that hunt at night, such as vipers. The third type of eye shape – horizontal – is seen only in a handful of tree snakes. Horizontal pupils offer wrap-around vision, helping snakes see in more directions at a single moment.

Smell and taste

Although snakes can take in smells directly through their nostrils like we do, they have an added bonus: they

can 'taste' smells too. By flicking its tongue in and out of its mouth, a snake can collect odour molecules from the air, before transferring them to a special odour receptor on the roof of the mouth called the Jacobson's organ (or, more formally, the vomeronasal organ). The forked nature of the snake's tongue helps this special organ ascertain the direction from which odours may be coming.

Feeling the heat

In addition to the sensory adaptations discussed on these pages, boas, pythons and pit vipers have evolved a genuine super-sense. Members of these species can detect and locate the heat signatures of potential prey. They do this with special heat pits, which are lined with cells containing thermoreceptors that are highly sensitive to slight changes in temperature.

Some snakes can detect temperature changes in their local environment as small as 0.2°C (0.36°F), meaning they can hunt 'cold-blooded' animals, like lizards and frogs. Their thermoreceptors can even pick up the heat trails left by fleeing animals as they brush through the undergrowth.

The heat pits are located on either side of the head, meaning that a hunting snake can 'feel' the direction from which heat signals are

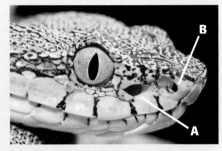

Above: Heat pits (A) are clearly visible as open holes on the side of the snake's head. In pit vipers, these occur between the eyes and the nostrils (B).

coming – in other words, the heat signals come in stereo. Indeed, so sensitive are these heat pits that they afford a snake a three-dimensional understanding of a given prey's exact location and distance. This means that, incredibly, even a blind rattlesnake can score a direct hit on prey 98 per cent of the time.

Hearing

The fact that snakes lack external ears doesn't stop them from having good hearing. At the back of the skull is a vestige of their ancestral ear anatomy: a small bone called the stapes, which sends vibrations to the inner ear that the brain then translates as sounds. Snakes are especially sensitive to vibrations that come through the ground, including low-frequency sounds that indicate the arrival of larger predators. The lower jaw acts as a conductor for these sounds.

Touch

For many years, scientists have discussed the significance of tiny pits (including so-called tubercles) on the bodies of snakes, particularly on the head. Many of these tubercles are closely aligned to regions with an abundance of nerve endings, so they must have an important role in helping snakes sense their surroundings. Undoubtedly, the pits help snakes move through their environment, but they may also play a part in detecting sexual smells (pheromones) and touch. At least one species of water snake can also detect light using these mysterious features.

Generous jaws

Above: While swallowing large prey (such as this deer) a large snake is open to attack from other large predators. For this reason, snakes often regurgitate prey if they are interrupted.

Below: Open wide! The powerful jaws of the African Rock Python (*Python sebae*) are capable of swallowing antelopes, warthogs and even hyenas.

The jaws of all snakes are famously flexible. However, the popular idea that they can be unhinged is perhaps a literal stretch too far. Instead, snake skulls are better imagined as being loosely articulated and held together with bands of elastic ligaments, which means that the jaws can be pulled back into place after being stretched. As a result, most snakes can eat prey many times their size, or perhaps more accurately, many times their girth. But swallowing prey takes time, particularly if the prey is large.

To help them swallow large food items, many snake species can move the left- and right-hand sides of their jaws independently. This means that they can 'walk' their jaws over a meal, ratcheting prey items deeper and deeper into the throat.

Swallowing large food items is a complex operation for snakes, not least because they are at their most vulnerable to predators when incapacitated in this way. For this reason, they often regurgitate prey if spooked or – as happens frequently – if suddenly interrupted by nosy humans or their pets.

Learn your scales

Left: Each snake species has its own unique scale patterns on the head. These scales offer protection from prey, parasites and the sun.

If you look closely at a snake, you will see an intricate mosaic of armour that offers both protection and durability. And if you look closely at a range of species, you will see that each has its own unique scale patterns and intricacies.

The purpose of scales

Undoubtedly, an important role of scales is to protect snakes from predators, as well as to afford them protection from any prey that fights back. But scales also offer protection from parasites and, vitally, from the drying effects of the sun. In addition, many scales have at least one or more spiky edges that help the snake to grip and push off from objects in order to move (see page 14), and scales are what give snakes their patterns and distinct markings. In this way, each and every scale is like a Swiss Army knife of adaptive endowments.

Types of scales

In general, snake scales come in three types. The dorsal scales run along the snake's uppermost side, and are often slightly keeled in the middle to provide added streamlining. Ventral scales are situated on the underside of the snake. These widened scales are incredibly smooth to reduce drag as the reptile moves along the ground.

The head scales often vary greatly between snake species. They are far more plate-like and armoured than the other two types, and slot together in a jigsaw-like pattern.

Snakes also have scales on their tail. Some species have evolved unique pointed scales on their tail that can be used to manipulate prey. In contrast, some burrowing snakes have evolved a blunt tail that can be used like a plug to protect them from digging predators trying to excavate them from their subterranean burrow.

Sloughing

Throughout their lives, snakes regularly shed their outer layer of skin as they grow, a process called sloughing. Often, sloughing begins with the snake rubbing its head against something sharp and then pulling the skin off by moving its body forward. The skin comes off intact, often in one piece, complete with tiny goggles – the scales that were once part of the snake's eyewear. If you come across a discarded snake skin, carefully take photos of it and (with permission if necessary) keep it somewhere safe. The scale patterns on the shed skin will tell you which species it belonged to, potentially helping local wildlife conservation initiatives.

Below: A discarded slough can make a memorable addition to a school's nature table or a talking point for an afterschool nature club.

Super-slim anatomy

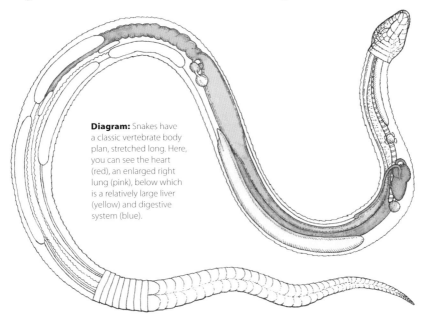

Diagram: Snakes have a classic vertebrate body plan, stretched long. Here, you can see the heart (red), an enlarged right lung (pink), below which is a relatively large liver (yellow) and digestive system (blue).

From the exterior, snakes are silent, stealthy predators, stretched to the upper limits of hunting prowess. But look inside at their internal anatomy, and you will discover that trade-offs have been made in their various organs as their long, thin body plan has evolved.

Most notable in this respect are the lungs, one of which is dramatically larger than the other. Apart from pythons and boas, all snakes use their right lung for respiration. The left lung is a tiny sac – a vestigial remain. In many water snakes, the right lung is especially enlarged, allowing it to be used a little like a buoyancy aid.

To allow them to continue breathing while their mouth is otherwise engaged in swallowing prey, snakes have a muscular windpipe (the glottis), which they can push out of the side of the mouth so that they don't suffocate while feeding. Snake digestion begins in the mouth. Here, special glands secrete a digestive saliva containing enzymes that help ready the prey for digestion once it is swallowed. In some snakes, these glands are highly

Below: A snake's glottis is highly mobile. Like a snorkel it can be pushed to one side while swallowing large prey. This allows the snake to keep breathing as it swallows.

Above: Like birds, crocodiles and lizards, snakes produce their waste through a single opening, the cloaca.

Below: Snake droppings are noticeably powdery. Sometimes the chalk-like remains of bones and teeth can be spotted.

modified and provide the vehicle for venom production. In these cases, fangs provide the mechanism for delivery.

From the mouth, food passes into the stomach, which is little more than an enlarged section of the gut. Here, the bulk of digestion takes place, before the food passes into the small and large intestines. The long, thin body shape of snakes puts a constraint on how much the intestines can coil around, meaning their digestive tract is notably shorter than in many other vertebrates. Droppings pass out of a single excretory hole, the cloaca, through which waste from the kidneys also passes. To save water, snakes excrete metabolic waste as uric acid, giving their urine a dry, paste-like consistency.

Like mammals, snakes fertilise their eggs internally. Males have a pair of copulatory organs called hemipenes, which function rather like the penis of a mammal. In some females, the evolutionary pressures of space saving have again taken their toll, leaving them with a dramatically reduced or absent left ovary.

Hunting styles

In nature, no two prey species are exactly the same. Some prey items, such as rodents, try to run away very quickly when grasped. Other prey items, such as fish, can be spiny. And yet others, including the Common Toad (*Bufo bufo*), may fight back with weapons such as poisonous skin. Because one strategy rarely fits all in nature, some snakes have adapted to preying on preferred species, which puts a fascinating context behind their hunting behaviour.

Above: Many of the poisonous defences of amphibians, including those of the Common Toad, have evolved as a deterrence to snakes.

Many snakes are active hunters of prey. These species spend much of their time seeking out potential burrows and cracks in which their prey may be hiding. Nocturnal snakes in particular excel at this hunting strategy. To assist them, these snakes follow fresh odour (or heat) trails or, in the case of snakes that hunt snails and slugs, even slime trails.

Other species prefer to lie still and wait for prey to come to them. Of these, some snakes, including rattlesnakes, seek out 'runs' – special corridors that small mammals use as a kind of highway between

Below: A rattlesnake striking at prey. In some species, their lunging jaws can move at a speed of more than 3m (almost 9ft) per second.

Egg-eaters

Arguably, the closest snakes have got to a more vegetarian way of life is in those species that regularly locate and eat eggs. Some snakes, particularly the African egg-eating snakes (genus *Dasypeltis*), depend almost completely on bird eggs for their diet. First, the snake engulfs the egg in its mouth. Then it forces the egg against special spines that protrude from the neck vertebrae. When the egg cracks against these spines, the snake swallows the nutritious contents before regurgitating the crushed eggshell in a neat little membranous package.

Above: Twelve species of true egg-eating snake are known. Each relies on tooth-like throat spines to crack open its prey.

feeding grounds. They position themselves next to these runs, often for hours on end, waiting for an opportune moment to strike. Another group of snakes attempts to attract prey to them. One example is the Spider-tailed Horned Viper (*Pseudocerastes urarachnoides*), which has modified scales on the tail region that resemble a small spider. By wiggling its tail theatrically, the viper can attract inquisitive (and hungry) birds before striking out from the undergrowth with fangs erect.

Among the most famous of the sit-and-wait specialists are the constrictors. These snakes are often large – their big, bulky, often coiled body provides a firm anchor point from which to strike with precision. The constrictors include some of the most camouflaged of all snakes, including the boas and pythons. They specialise in

immobilising their prey quickly, stopping them from fighting back so that the process of swallowing can begin in earnest. Constrictors often kill their prey by 'throwing' their coils over their quarry, before tightening the coils through specialised muscle contractions. This squeezing action restricts the prey's ability to take a breath, as well as cutting off blood circulation. Only the biggest and strongest prey items may have the power to resist.

Above: A Spider-tailed Horned Viper complete with tail lure. The viper waves this tail conspicuously to catch the interest of arachnid-eating birds, striking when they get too close.

Below: Constricting snakes monitor the heartbeat of their prey to ascertain whether they are still alive and capable of escaping.

Meet the Residents

Of the 50 or so snake species native to Europe, the UK is home to just three – the Adder, the Smooth Snake and the Barred Grass Snake (previously called the Grass Snake). Britain is also home to a handful of non-native snake species, which are included in this section along with one native legless lizard, the Slow Worm, that is often confused for a snake. While we may lack species numbers, our snakes more than make up for this with their charisma, flair, endeavour and diversity of hunting styles.

Barred Grass Snake

In terms of length, the Barred Grass Snake (*Natrix helvetica*) far outstrips other British snakes. Regularly growing to 90cm (35in) or more, this athletic predator has no problems swallowing large prey such as frogs, fish and even birds. Compared to the other British snake species, the Barred Grass Snake is widespread, being found throughout England and Wales, and even into southern Scotland. It is often associated with fresh waters, and is mercifully easy to identify courtesy of a yellow or orange band of colour behind the neck.

Adult identification

The key feature of the Barred Grass Snake is its distinct orange or yellow collar, flanked at the rear by a crescent-like black marking. This patterning is seen on nearly all individuals, both young and old. The 'bars' in the common name refer to the black lines along the flanks of the body. Again, these are common to most individuals, although not all.

While an exceptional Barred Grass Snake may reach up to 150cm (60in) in length, most individuals seen in the wild rarely exceed 90cm (35in). In fact, many Barred Grass Snake encounters involve younger individuals that are perhaps only 50cm (20in) or so in length.

Above: The obvious yellow-and-black collar of the Barred Grass Snake is prominent and easy to spot. Nearly all individuals possess this marking.

Opposite: Older Barred Grass Snakes, like the one opposite, have a notably wider skull – this allows them to catch and swallow large prey items, such as adult frogs and toads.

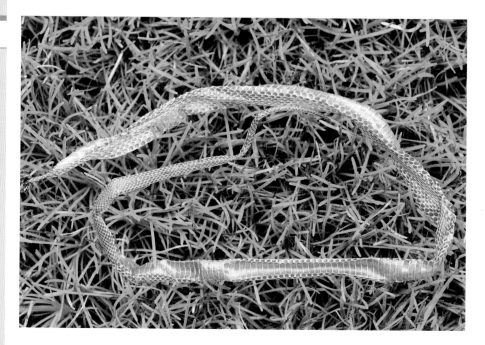

Above: A raised ridge runs along the middle of keeled scales. These scales scatter light in a way that can make the scales look less shiny.

Slough ID

The Barred Grass Snake sheds its skin as a single piece, on which the scale patterns and colours are barely discernible. The scales on the uppermost side are notable for being keeled, while those on the flanks are less keeled.

Habitats and distribution

On the whole, if you see a snake swimming in England or Wales, it's highly likely to be the Barred Grass. The species is known to occupy woodlands, commons and gardens near fresh waters such as rivers, ponds and lakes. Individuals do stray from these areas, however, particularly when searching for egg-laying sites, which include garden compost heaps and rotting piles of manure.

In recent years, confirmed sightings of Barred Grass Snakes have been made in Dumfries and Galloway, making the species a new native of Scotland as well as indigenous to England and Wales. It is also present on the island of Jersey.

A year in the life

The Barred Grass Snake is slightly more limited by climate than our other widespread snake species, the Adder. Southerly populations may emerge from their winter slumber in March, whereas more northerly populations may rouse themselves later, when the warm sun really begins to kick in. In the early days of spring, the Barred Grass

Above: The Barred Grass Snake is a powerful swimmer. When disturbed in water it dives down to hide among pondweeds, holding its breath until the threat moves on.

What's in a name?

The Grass Snake (*Natrix natrix*) was until very recently considered to be a single widespread species found throughout Europe. However, in 2017 this state of affairs changed. Scientists undertaking DNA analyses of Grass Snakes across Europe discovered that what was once considered to be one species was actually two distinct species: a westerly species, to which they gave the common name Barred Grass Snake and the scientific name *Natrix helvetica*; and a more easterly species, which retained the names Grass Snake and *Natrix natrix*. Almost overnight, the species that until then had been known in Britain as the Grass Snake received a name change, and following this it is referred to as the Barred Grass Snake on these pages.

Right: After coming out of its winter slumber, the Barred Grass Snake spends much of its time basking in the springtime sun. This is a great time to spot reptiles.

Below: Barred Grass Snakes often choose the same local spot (hibernacula) during the winter months. In the early days of spring, they may emerge and bask together.

Snake is likely to spend much of its time sunbathing and readying itself for mating, which begins in April or May.

Mating is a suitably combative affair, with males squirming and wrestling with one another to gain access to females. Eggs are laid in June once the female has sought out a suitably warm and protected location. Rotting vegetation provides ideal real estate for the female Barred Grass Snake – the heat produced from the surrounding decomposition acts as a helpful incubator for the eggs. This is the only native British snake that lays shelled eggs, which resemble miniature ping-pong balls in their colour and texture. Within three months, the eggs crack open and the tiny newborns emerge, each little more than the size and stature of a pencil.

Barred Grass Snakes begin to head toward their overwintering areas in October, when the days become short and frosts more common. Like other snakes, they often hibernate in disused burrows, sometimes alongside other reptile species such as the Adder and Slow Worm.

Getting acquainted

The idea that there were snakes living among us in Britain brought a fantastical element to my early naturalist pursuits. It excited me immensely that such enduring, prehistoric-looking predators could exist in and around our neighbourhoods. On hot days, I would traipse along canal towpaths searching for snakes basking on the banks. Yet, for me, that sacred encounter never came. It wasn't until much later in life that I realised there are some key pointers if you want to see and study native Barred Grass Snakes.

The first thing to remember is timing. Barred Grass Snakes often bask early in the day and begin to hunt only when their temperature is optimised. Therefore, walking quietly along sunny banks of ponds, lakes and canals in the cooler early mornings throughout May and June can improve your chances of seeing this species immensely.

Second, think like a Barred Grass Snake. These reptiles do not like to be disturbed and rarely bask in the open, for fear of snake predators like Red Foxes (*Vulpes vulpes*) and birds of prey. For this reason, the best potential

Above: A newly hatched Barred Grass Snake next to its marble-sized egg. These early days are among their most dangerous, when predators of baby snakes (including birds) abound.

Above: The evolutionary reason for the Barred Grass Snake's collar is not yet known. It may be that the collar helps to camouflage the snake from predators as it moves on the water's surface.

basking spot will have beside it an escape route in case the snake is spooked. Often, these reptile 'panic rooms' include a dense patch of scrub or long grass. Barred Grass Snakes particularly like to rest underneath warm objects to benefit from their heat while remaining safely tucked away. These items (often called refugia) include logs and tree roots as well as human-made objects such as patio slabs, compost heaps, bits of old carpet and discarded scrap metal. Look for Barred Grass Snakes hiding under or sunbathing on top of these refugia in early spring.

It is extremely important to keep your distance from snakes when observing them. A disturbed snake moves on quickly, a behaviour that wastes its precious energy and can interrupt hunting. It can also alert predators, such as birds of prey, to the reptile's whereabouts. Observe snakes such as these with binoculars, ideally, and enjoy them as you would a rare and elusive bird.

Conservation

Once common and widespread throughout most of lowland England and Wales, the Barred Grass Snake has gone the way of many reptiles and amphibians in recent years, by becoming isolated into smaller and smaller patches of suitable habitats across much of its range.

Record-breakers

Above: Few predators in Britain are able to tackle large snakes. Once they reach a certain age, therefore, their survival chances are likely to increase dramatically.

As is often the way with snakes, controversy abounds in Britain over the largest Barred Grass Snake specimens ever caught. While seasoned snake-watchers may have had occasional brushes with large individual Barred Grass Snakes that are perhaps 150cm (60in) or so long, the record-holder is surely that claimed by forester John Wood. In the 1950s, he reported measuring an individual snake in the New Forest at 180cm (78in) in length.

Of the hundreds of encounters I have had with wild Barred Grass Snakes, I have come across snakes approaching the lengths recorded by Wood only a few times. On one occasion, however, while lifting up refugia during a reptile survey, I came face to face with an extremely large, 150cm (60in)-long female, who opted to do nothing by way of retreat. It was almost as if her immense size meant she did not see me as a potential threat – and she was right. We simply watched one another in a mutually respective trance, before it was time for me to move on. Memorable wildlife encounters don't get better than this.

Locally, these fragmented populations are much more vulnerable to extinction – through loss of egg-laying sites or hibernation places, for instance, or through disturbance by dog walkers or those who persecute snakes regardless of the legal penalties. Understanding and formally recording snake sightings has become an important tool in protecting and securing the future of isolated local populations of the Barred Grass Snake. Each and every sighting matters.

Above: Before saving snakes, scientists need to know where they are. Here, an ecologist is looking under a specially placed 'refugia' to determine the absence or presence of the Barred Grass Snake.

Wetland nature reserves are an important stronghold for the Barred Grass Snake, particularly those within former brick and gravel pits. Managed correctly, these nature reserves can be home to populations in the hundreds or even thousands, potentially providing healthy populations capable of 'reseeding' nearby urban and suburban environments.

Gardens also offer a potential habitat resource to some populations of Barred Grass Snakes, particularly larger gardens with wildlife ponds in which frogs and other amphibians flourish. For this reason, many wildlife conservation organisations have promoted positive messages about snakes and raised awareness about declines in recent decades to encourage garden owners to celebrate their neighbourhood snakes. However, there is still some work to do in raising the profile of the Barred Grass Snake if we are to convince everyone of the species' ecological value and inherent beauty. Books like this one will, I hope, help in this respect.

Left: Farms can be important habitats for snakes. In particular, piles of hay or manure provide important egg-laying sites.

Below: Garden wildlife ponds – a haven for frogs – could become an important resource to suburban populations of Barred Grass Snakes.

The Adder

Few animals conjure up quite as much fear as the Adder (*Vipera berus*), with its venomous bite. But Adders are about much more than their bite. For a start, they are beautiful animals – in some lights, their scale patterns are almost feathery. Their behaviours are incredibly complicated and complex, and the snakes are also enduring and tough – they are one of the world's hardiest vipers. Yet, in Britain at least, Adders are not as widespread as they once were. Over time, their habitats have been pieced off into smaller and smaller chunks, and the snakes have found themselves increasingly rubbing up against people and their pets. Of all of the UK's reptiles, the Adder is of deepest concern to wildlife conservationists.

Adult identification

The Adder is a much shorter, stockier snake than the Barred Grass Snake. Although large males reach 55cm (22in) in length and large females 70cm (28in), most Adders are smaller. On the snake's dorsal surface is a distinct dark zigzag pattern that runs down the total length of the body. The eyes are red or orange, and the pupil is notably vertical compared to that of our other native snakes. Male and female Adders can often be

Below: A basking Adder. Note the 'lightning bolt' pattern that runs down the back and the prominent red or orange eyes.

told apart by their coloration: males have a defined black zigzag against a white (almost silver) body colour, whereas females are copper in colour with a brown zigzag pattern.

Above: The zigzag pattern of the male Adder contrasts more sharply with its background. Females often look notably bronze in colour.

In some parts of Britain, populations of completely black (melanistic) Adders are known. This genetic anomaly gives these individuals a unique kind of beauty – one that has been revered by generations of local snake-watchers.

Slough ID

Adder sloughs are fairly easy to identify. Holding the skin up to the light should expose the familiar zigzag pattern, and if you look more closely you will see that the scales are obviously keeled. In spring, Adder sloughs are often found near hibernation sites.

Habitats and distribution

The Adder is particularly associated with commons, moorland, heathland and chalk downland. It also exists in patchy populations in some open woodlands. Here, in spring, woodland rides provide valuable basking sites. Populations can also occur on railway and road embankments near good reptile sites. In some cases, these embankments provide important 'bridges' between

Above: The charismatic red or orange eyes – with prominent vertical pupil – is clearly observable in all Adders.

two ancestral populations of Adders. Unless your garden happens to back onto such a place, you are unlikely to see the species close to your home.

The Adder once lived across much of Britain, but sadly, persecution and habitat changes have pushed it to the very edge of extinction in most parts. In much of central England, for instance, the majority of Adders occur in only scattered and discrete populations. However, it's not all bad news. In some isolated places where small mammals abound, Adders can thrive in numbers that might come close to reflecting their former glory. These sites include the Pembrokeshire coast (where many Adders live among the cliffs), parts of Dumfries and Galloway, and in the New Forest.

Right: Being excellent places for many rodents, brownfield sites – such as former brick and clay pits – can become important strongholds for local populations of Adders.

A year in the life

The Adder awakens from its winter slumber earlier than other British snakes. Often emerging as early as February, the snakes head to nearby south-facing basking points, where they remain largely motionless for hours at a time. This is an important time for both male and female Adders, as the warm sun helps them produce eggs and sperm for the reproductive festivities to follow.

Adder mating begins about a month after the snakes emerge from hibernation. Males chase down the scent of receptive females, sometimes meeting rivals in the process. Duels between males are frequent, sometimes leading to the so-called 'Dance of the Adders'. Here, male Adders rise up against one another, twisting and tussling in a spectacle that is half arm-wrestle and half ribbon dance. Inevitably, the largest male proves most powerful in these encounters, forcing the smaller male to retreat. The female will mate with the winner, who then remains nearby to guard her from other male advances. After mating, Adders become largely solitary animals.

Above: The famous 'Dance of the Adders'. Here two males face off against one another to win the interests of a nearby female Adder.

Most Adders hunt using a 'sit-and-wait' strategy. As in other venomous snakes, the Adder's fangs are laced with a fast-acting venom that incapacitates its victim, stopping it from running too far away after it is bitten. The Adder then follows the scent trail to gather up its now immobilised prey.

Left: The Adder's fangs are its venom-delivery system. But venom is costly to produce and so Adders must use their bites sparingly.

Above: A female Adder with offspring. Normally the young move away from their mother within a few hours of being born.

Young Adders are born live, in numbers of up to 20 or so, each encased in a membranous sac rather than a hard shell (see page 79). Often, these young can be spotted in late August. Each is little more than 16cm (6in) in length at birth, yet the charismatic zigzag pattern can clearly be seen.

The Adder normally returns to the same hibernation site each year, sometimes hibernating alongside other reptiles such as Slow Worms and Common Lizards (*Zootoca vivipara*). Most Adders have normally retreated to overwintering sites by about November, although at some sites individuals may be spotted in December and even January.

Staying safe

According to National Health Service (NHS) figures, about a hundred people are bitten by Adders each year. Although many of these bites are painful, none (at the time of print) has proved fatal to humans in more than 40 years.

Contrary to certain media reports, the Adder uses its venomous bite on humans only in self-defence, particularly if cornered or accidentally trodden on. Sometimes, people who get too close are bitten by an Adder – for example, when trying to pick up a snake or

What to do if you are bitten

If you think that you or your pet has been bitten by an Adder, seek medical advice immediately. At the same time, the NHS advises snakebite victims to follow the dos and don'ts below:

Do

- Call for help – ring 999 or ask someone to drive you to a hospital.
- Remain calm – take deep breaths.
- Try to remember the size and colour of the snake.
- Rest up while you're waiting for help. Keep the bitten area of your body still to prevent the venom from spreading to other parts.
- Remove watches, rings and other jewellery, as a bitten limb is likely to swell.

Don't

- Try to suck out the venom.
- Try to cut the venom out or make it bleed.
- Apply ice, heat or chemicals to the wound.
- Allow yourself to be alone at any point.
- Apply pressure bands, tourniquets or ligatures to the affected limb to stop the spread of venom. This can make the swelling worse and complicate recovery.
- Try and kill the snake. This is illegal under UK law.

take a close-up photograph. This behaviour is not good practice, and the distress it causes snakes could actually be detrimental to their local conservation.

Before unleashing a defensive bite, most Adders will give lots of warning – they hiss and draw their head and neck back over their body, delivering mock strikes. Three ways to all but eliminate the risk of you or your pet being bitten are keeping to paths, wearing correct footwear and keeping dogs on a lead at Adder sites.

Producing venom is costly to Adders. It is therefore in their best interest to conserve venom, and they do this by inflicting so-called 'dry bites' when threatened. These bites can puncture the skin of a would-be attacker, but only a small amount of venom (or sometimes none at all) is delivered. Thankfully, approximately 70 per cent of Adder bites on humans are dry bites.

ADDERS
YOU MAY FIND ADDERS
BASKING IN THE GRASS
AND AMONGST THE HEATHER

THEY ARE A LEGALLY PROTECTED SPECIES
PLEASE DO NOT DISTURB THEM

KEEP DOGS UNDER CLOSE SUPERVISION
SEEK IMMEDIATE MEDICAL OR VETERINARY HELP IF BITTEN

Walberswick Common Lands Charity

Above: Many nature reserves choose to erect signs to warn walkers about the presence of Adders. Crucially, these signs remind people of the protected status of snakes in Britain.

Getting acquainted

Some well-managed heathland sites in Britain do have healthy Adder populations. This means that, during early spring, interested observers have the opportunity to watch these snakes basking. Adders often position themselves in south-facing sunny 'thrones' between small tussocks of grass and clumps of dried Bracken

Above: Reptile walks – where the public are given a chance to spot reptiles – are a useful public relations tool for snakes. Just be mindful of disturbing basking snakes.

Right: The distinctive zigzag pattern of the Adder is likely to have evolved to camouflage within Bracken, a common plant found within its distribution.

(*Pteridium aquilinum*). Almost certainly, their charismatic camouflage is an adaptation that mimics the zigzag edges of the fern fronds.

Fun and rewarding as snake-watching can be, remember that spring is an important time for Adders, particularly as they build up reserves for the mating season that follows. For this reason, do not get close to an Adder or intentionally disturb one while it is basking. Like birds, Adders can be photographed very well through a scope or a camera with a high-quality zoom. It is irresponsible to take close-ups with a mobile-phone camera, as you will be disturbing the snakes and also putting yourself at risk. And remember that not all wildlife encounters look better through a lens. I'd argue that this is especially the case with Adders.

Adders often choose to bask in early spring near their hibernation sites, in what are often known locally as 'Adder banks'. If you do see Adder banks at this time of year, inform the landowners as they may not know about them. This can help protect hibernation sites from being managed in a way that inadvertently damages the snakes.

Conservation

Being resident in a wide array of habitats, Adders likely once thrived throughout many parts of mainland Britain. Today, however, the scale of their decline has made them a most urgent priority for conservationists – arguably more so than any other UK reptile species. Across Britain, and particularly in central parts of England, Adder losses are spectacularly depressing. In recent decades, it's likely that the Adder has become virtually extinct in a number of English counties, including Hertfordshire, Nottinghamshire, Warwickshire, Cambridgeshire, Greater London, Lancashire and Oxfordshire. Any Adders that remain in these counties are likely to exist in small, restricted populations that are particularly susceptible to local extinction.

Above: Adders are far harder to spot than they were a century ago. A suite of human-caused threats to their populations exists, which conservationists are working hard to tackle.

Put bluntly, the Adder's current distribution is a shadow of what it once was. Thankfully, the scale of the loss is galvanising conservationists to take action now, before it's too late.

Smooth Snake

For many years, the Smooth Snake (*Coronella austriaca*) was considered to be a peculiar and highly secretive subspecies of the Grass Snake. Only in 1859 did scientists realise that it was actually an entirely separate species, with its own temperament, hunting style and rather picky habitat requirements. Today, the Smooth Snake is a resident of a limited number of specially managed nature reserves in southern England. However, continued successful reintroduction projects could be a key part of this snake's future in Britain.

Adult identification

Reaching a maximum length of 45–55cm (18–22in) and with a slim build, the Smooth Snake is notably more diminutive than other British snakes. Adults of both sexes have a dark heart shape on the top of the head, which can help with identification. Adults also possess a black bar that extends from the nose and across the (often red) eye. Unlike Barred Grass Snakes (with which they can be confused), Smooth Snakes lack a yellow collar. Female

Below: A Smooth Snake. Note the heart-shaped pattern on the back of the head and the lack of a yellow-black collar as seen in the Barred Grass Snake.

Smooth Snakes can often be told apart from males by their greyer underside, and they also lack the slight ginger coloration males are said to have on the underside of their head.

Slough ID

The Smooth Snake gets its name from its streamlined scales, which have no obvious keel. Its sloughs are more fragile than those of either the Adder or Barred Grass Snake and are sometimes found in and around their refuges.

Left: The slough of a Smooth Snake. Note that each of the scales looks like a featureless window – there is no 'keel' as seen in the dorsal scales of other British snakes.

Habitats and distribution

Although Smooth Snakes may once have thrived throughout the extensive heathlands of southern England, their distribution today is highly fragmented into distinct, well-managed chunks of heathland, particularly in parts of Dorset. In Hampshire, the species occurs in dense parts of the New Forest and in parts of the Avon Valley. It is also found on a number of special nature reserves in Surrey, and there are possible historical reports from Cornwall, Berkshire and Wiltshire. Surrey, West Sussex and Devon are now home to reintroduced populations that are being carefully monitored and managed. There may be other small populations of Smooth Snakes out there, currently unrecorded, and protecting these hidden sites could save the species from a further downward slide.

Above: The bronze or brown coloration of the Smooth Snake can mean it is occasionally confused with the Adder. Note the lack of zigzag markings down the back and the circle-shaped pupil.

A year in the life

In most years, the Smooth Snake emerges from its winter quarters on warm days in March. Like other British snakes, it spends these early days mostly basking in the sun. The Smooth Snake can be very hard to spot while it is basking, partly because of its camouflage, which matches the colours of the surrounding dry heathers, but also because (unlike other British snakes) it likes to wrap itself around twigs and dense branches.

The Smooth Snake's breeding season begins in May. Little is known about the intricacies of its reproductive behaviour, most of which takes place hidden among the shadows of the undergrowth. The young normally appear in August or September. Just like the Adder, young Smooth Snakes are born live and measure about the length of a pencil (13–15cm/5–6in).

The summer months are important to Smooth Snakes. Primarily being predators of small reptiles, which flourish in these warmer seasons, they take the opportunity to put on weight before autumn arrives. Most Smooth Snakes will have moved to hibernation sites by early October.

Conservation

The popular maxim that the squeaky wheel gets the oil applies well to Britain's wildlife. Many would argue that the big, showy creatures with doe eyes and oodles of charisma (and, dare I say it, fur and feathers) have traditionally captured the hearts and hence interest of the public, whereas the wildlife species that prefer to skulk in the shadows struggle to get noticed. Smooth Snakes are one such example. Being patchily distributed, highly secretive and specialising in heathland microhabitats that human eyes find hard to penetrate, this species was almost totally lost without anyone really noticing. What was once a snake that thrived throughout the southern heathlands of England almost disappeared without a single hiss being heard by anyone until it was almost

Seeing the Smooth Snake

Most people living in Britain are unlikely ever to see a Smooth Snake, so secretive and patchily distributed has this species become. However, organisations such as The Amphibian and Reptile Conservation Trust occasionally run special guided walks to see rare reptiles such as this. Although you may not see a Smooth Snake every time, you're in with a good chance of spotting reptiles of some sort – including Common Lizards, Slow Worms and Barred Grass Snakes. Details of these events are posted on the Trust's website (see page 125). Remember that the Smooth Snake and the habitats the species depends upon are highly protected under UK law. Many activities involving the Smooth Snake, including research projects, require a special government licence.

too late. The decline of the species wasn't due to direct human interference, but rather the loss of the heathlands with which they associate. Today, only one-sixth of the lowland heathland that existed 200 years ago is still with us, and these remaining patches of land face a volley of threats and pressures, including air pollution, recreational disturbance and human-caused heathland fires.

Undoubtedly, the fact that the Smooth Snake still remains in Britain is down to the hard work of conservationists and scientists and the organisations they represent. Through their passionate voices, the Smooth Snake is finally getting the attention it deserves. The squeaky wheel is finally being greased, although there is much work still to do.

Above: Because Smooth Snake populations are often very isolated, conservation groups have to keep a watchful eye on local threats.

To save the Smooth Snake from extinction in Britain, scientists have pursued a three-pronged approach. In addition to protecting and monitoring existing sites and populations, and enhancing these sites for reptiles, scientists have also successfully reintroduced Smooth Snakes to other parts of their former range. In some cases, this has been through the removal of individuals from one habitat and their careful release elsewhere, while in other cases, carefully reared captive-bred Smooth Snakes have been used. The wildlife group The Amphibian and Reptile Conservation Trust, working alongside other organisations such as the RSPB and the National Trust (see page 126), has made impressive headway in reintroducing Smooth Snakes.

Snake-like lizards – the Slow Worm

Above: The scales of the Slow Worm are incredibly shiny and smooth. This helps reduce drag as it burrows through the undergrowth.

Below: Up close, the Slow Worm retains familiar lizard features including a more rounded head, earholes and eyes that blink.

Some modern-day lizard families have evolved progressively smaller limbs, just as the ancestors of snakes did during the early days of dinosaurs, more than 150 million years ago (see pages 6–7). These include the lizards of the genus *Anguis*, sometimes called the blindworms. Britain is home to a particularly endearing blindworm species, the Slow Worm (*Anguis fragilis*), a widespread and rather charming resident of England, Wales and Scotland. Slow Worms are frequently misidentified by members of the public as snakes, and for this reason they are included here.

Identification

The Slow Worm exposes its lizard credentials in three obvious ways. First, in common with all lizards (and unlike snakes), it can blink. Second, it has tiny earholes, which snakes lack. Third, like many lizards, the Slow Worm can shed its tail. Look more closely at Slow Worms and

some other differences become apparent. For a start, all
Slow Worms are more cumbersome than snakes, hence
their name. They also lack the highly evolved muscular
packaging and modified skeleton that snakes possess. And
their smooth, densely packed scales give them a shiny,
reflective texture in comparison to snakes. In some lights,
the skin of a Slow Worm can almost reflect like a piece of
shiny jewellery, especially in the case of newborns. The
Slow Worm measures as little as 8cm (3in) when born
and up to 50cm (20in) as an adult.

The Slow Worm often sheds its skin hidden from view
underneath logs and in other hiding places. Unlike snakes,
its shed skin forms a dense 'ring' that comes off rather like a
sock being rolled down a human ankle. Close observation of
Slow Worm sloughs reveals their tiny smooth scales.

Habitats and distribution

The Slow Worm is far more at home in gardens than any
other UK reptile. In fact, some Slow Worm populations
exist in the most urban of environments, where they
move through gardens and allotments, or along road and
rail embankments. As well as thriving
in gardens, the Slow Worm is also a
resident of many open woodlands,
hedgerows, heathlands and even sea
cliffs. Although this legless lizard can
become locally very common, many
populations are isolated from one
another and can be very patchy.

To find out if you have Slow Worms
in your garden, try dotting some old
pieces of carpet in sunlit spots where
invertebrates gather, such as compost
heaps or bark-covered soils. Carefully lift
up these potential reptile haunts in spring and summer,
and you could expose one or two of these wonderfully
charismatic members of our native reptile fauna.

Like our native snakes, the Slow Worm is protected by
law from harm or intentional injury. This is partly to guard
this important snake-like lizard from an equally unjust
and unfair persecution.

Above: Slow Worms can live in
dense populations. If there is plenty
of food around, such as ants or
slugs, they may regularly huddle
together under the same slab or
old bit of carpet.

A year in the life

Right: Seasoned watchers of Slow Worms may sometimes get to see them mating. Here, a male grasps a female in his jaws while copulation takes place. Often females possess scars as a result of this curious mating behaviour.

Like snakes, the Slow Worm emerges from its underground slumber in March and proceeds to bask (sometimes in full view) in nearby sunny spots in preparation for mating in the months that follow. Mating itself largely takes place hidden from human view and so remains understudied. Males regularly engage in competitive bouts with one another during this period. Once a female is secured, the male will grab her with his jaws to arrange both of their bodies in such a way that copulation can be successfully achieved. Later in the summer, the abdomen of adult female Slow Worms looks notably swollen – within their bodies may be 25 or more tiny lizard offspring, which are born live in the late summer. The newborn Slow Worms look very fragile, and at this early stage many are preyed upon by birds, including Pheasants (*Phasianus colchicus*).

Below: Woodlice are also occasionally eaten by Slow Worms. This – along with all the slugs and snails – is another reason why Slow Worms are so at home in our gardens.

In the summer, the Slow Worm rarely needs to bask. In fact, most adult Slow Worms spend their time hidden in thick undergrowth or underneath patio slabs feasting upon ant nests. Aside from ants and beetles, Slow Worms also favour Netted Slugs (*Deroceras reticulatum*), the pale grey slugs commonly found in gardens.

Non-native snakes

Intentional or accidental releases of snakes have occurred in the past in Britain and occasionally occur today, even though the intentional release of non-native species is illegal. Thankfully, these snakes are nearly always non-venomous species that are not likely to harm people or pets. Three of the most likely non-native snake species you may, in the most unlikely of circumstances, come across in Britain are described below.

Aesculapian Snake

Two populations of this species are known in Britain, both founded by snakes that escaped from local zoos, one in central London and one in north Wales. Superficially, young Aesculapian Snakes (*Zamenis longissimus*) resemble Barred Grass Snakes. They have the same yellow collar, although juvenile Aesculapian Snakes have a black patch underneath each eye that looks (to my mind) a little like mascara that has run down the face. Adults are a uniform olive-yellow or brownish-green colour, sometimes with white freckles on the uppermost side of the body.

This is one of the largest snake species in Europe, with individuals occasionally reaching more than 2m (6ft) in length. In Britain, however, individuals are notably smaller than this due to the cooler climate.

Above: A beautiful Aesculapian Snake gets ready to strike. Note the 'mascara smudges' around the eyes. Most individuals are uniform in colour with few obvious markings.

Both populations of Aesculapian Snakes known in Britain are thought to be limited in their capacity to reproduce, hence conservationists do not consider the species to be invasive. While feeding opportunities abound in London in the form of rats and urban birds, the population here may be limited by the availability of suitable egg-laying sites. In time, both of these British populations may decline toward eventual extinction.

Above: Many escaped Corn Snakes probably end up being hunted by urban foxes and cats. The ones that survive likely perish over the winter months.

Corn Snake

Compared to many snakes, the non-venomous Corn Snake (*Pantherophis guttatus*) has a docile nature and a reluctance to bite in self-defence. This, combined with its large size (up to 180cm/70in long) and attractive patterns, has seen the species become a hugely popular pet in recent years. Sadly, because Corn Snakes – like all pet snakes – have a canny knack for escaping their confines, they regularly find their way into the wild. The native range of the Corn Snake is the southwestern United States, which is far warmer than Britain. As a result, most pet escapees end up seeking shelter in woodpiles or underneath garden sheds during anything but the hottest weather. Most are likely to be killed off during wintry cold snaps, for which they are ill adapted.

The Corn Snake is the non-native snake most likely to be spotted in Britain, often in highly urban or suburban localities. It is notably more slender than other snakes of a similar size, and its predominantly orange body is interspersed with patches of deep red or brown, outlined in black.

Pythons and boas

Very rarely, larger snake species escape their owners and live temporarily in the wild, particularly during spells of hot weather. Thankfully, these docile snakes are not venomous, although they can be predators of native wildlife, particularly rats and urban birds. In the summer of 2018, for instance, a busy high street in London was brought to a momentary standstill while a Boa Constrictor (*Boa constrictor*) in the middle of the pavement swallowed a pigeon. Thankfully, the snake in question was removed safely by officers from the Royal Society for the Protection of Cruelty to Animals (RSPCA).

Although the vast majority of reptile owners are responsible carers for their pets, occasional cases like these symbolise the problems that non-native species can inflict on local native wildlife when things go wrong.

Above: Snake-owners have a responsibility to safely look after their pet. One way to do this is to regularly check snake-housing for cracks through which snakes might squeeze. Releasing snakes intentionally into the wild is illegal.

What to do if you see a non-native snake

In most instances, the RSPCA and SSPCA in Scotland will be able to retrieve and rehome or relocate a pet snake lost by its owner. However, all observations of non-native species in the UK have value to conservationists, who are eager to monitor and understand the reasons why illegal releases occur and the impact they may have on native species, including through the spread of non-native reptile diseases. The public are the eyes of the nation! The Amphibian and Reptile Conservation Trust runs an important project called Alien Encounters, which monitors and records observations of non-native amphibians and reptiles in the wild. To find out more or to report sightings, visit alienencounters.narrs.org.uk.

A Day in the Life

To survive and prosper, Britain's snakes must seek out the sweet spot between three competing priorities. First, they must maintain their body at an optimal temperature within which they can move, hunt and digest. Second, they must locate the microhabitats in which their prey hides using a suite of sensory skills. And third, they must stay out of potential predators' sight at all times. Thankfully, snakes have an impressive array of tricks to help them manage this impossible-seeming challenge.

Heating up

The first behaviour most snakes must engage in each day is a spot of basking to help them warm up. The time of day at which British snakes do this largely depends on the weather. In late spring and summer, the Adder – our most cold-tolerant snake – often emerges from its sleeping quarters when the temperature reaches about 10°C (50°F). Our other native species – the Barred Grass Snake and Smooth Snake – undertake little activity until the air temperature reaches a slightly balmier 15°C (60°F). In the warmer spells during spring and summer, when the sun also rises earlier, most British reptiles have done the vast majority of their basking by as early as 06:00 hrs or earlier. On these days, by about the time most humans are on their first cup of coffee, snakes may already be moving through the undergrowth looking for hunting opportunities.

Opposite: Basking snakes (such as this Adder) have to weigh up the benefits of basking in the open against the threat of being spotted by predators such as birds of prey.

Snake activity isn't regulated and impeded only by cold temperatures such as on cool mornings. Just as snakes can be too cold to be active, they can also become too hot. During some summers, for instance, when there are long spells of intensely warm temperatures that rarely fall below 25°C (77°F), snakes often slow their activities and find somewhere cool to hide for days or even weeks on end. One obvious benefit of this behaviour (called aestivation) may be to conserve water.

Above: Snakes of all species are incredibly sensitive to heat being given off by natural objects, such as exposed patches of grass or sun-drenched rocks that continue to radiate their heat late into the evening.

Above: By coiling around itself in a circle shape, the Adder can turn its body into a solar panel capable of gathering lots of the sun's rays while retaining as much body heat as possible.

So, why do snakes operate only within such strict margins of temperature? Why have they evolved to become so picky? Undoubtedly, one important reason is that snakes are unable to modify their own body temperature like we can and so need to optimise their bodily functions – particularly those (like digestion) that rely on temperature-sensitive enzymes. Another reason snakes are active only at specific temperature ranges is to avoid the likelihood that they are caught out in cold or hot temperatures which restrict their ability to defend themselves against predators. Hot temperatures in particular can be dangerous for snakes, since their nerve and muscle cells lose their ability to coordinate effectively in such conditions, potentially leading to death by suffocation.

Night swimming

While surveying for newts via torchlight, many naturalists (including myself) have reported seeing the occasional Barred Grass Snake hunting in ponds during warm nights in early summer. In my experience, this occurs on nights during successive runs of hot days – the classic summer heatwave, when the temperature at night is too uncomfortable for sleep. Although these snakes are clearly visual hunters, such encounters reveal just how highly tuned their other sensory skills must be. Quite how often this nocturnal hunting behaviour occurs in Barred Grass Snakes, and whether the reptiles sleep at this time of year and for how long, remains largely unknown. It may be that the climate crisis alters the behaviour of this species in Britain and that night-time sightings become more common.

Bask masters

To most passers-by, a single snake basking in the spring sunshine is very much like any other, but this is emphatically not the case. For each and every snake has made careful adjustments to its body shape and basking position in order to milk as much of the sun's energy as possible. Basking is a complicated business for snakes, and their basking styles often change throughout the day. Below is a guide to the different basking styles employed by British snakes.

Panelling

Many snakes (and lizards) increase the surface area of their bodies when facing the sun during basking. Often, they can be seen to spread their ribs wide and become noticeably flattened, forming something akin to a reptilian solar panel.

Below: While basking, many Adders keep a watchful eye on the sky. All snakes are capable of responding to passing shadows, particularly those of birds of prey.

Tilting

In early spring, Adders in particular favour sloping, south-facing banks quite near their hibernation sites for basking. By lying on a tilt facing the sun, almost as if they are on sunlounger, they absorb the solar rays uniformly across their body.

Hot-housing

Many snakes make use of objects that absorb and hold on to the sun's energy to warm themselves up. These objects are often man-made, such as old pieces of corrugated metal, sheets of panelling or disused carpet. As well as basking on top of (or even underneath) these items, snakes use them as handy places to hide or find prey. At many sites, these artificial refuges provide a handy focal point at which local reptile populations can be studied.

Above: Basking sites are often next to thick undergrowth and can be used by more than one individual at once. Note the all-black Adder in this image – a rare 'melanistic' form occasionally spotted in the wild.

Grouping

Both the Barred Grass Snake and the Adder often bask in flattened areas between tufts of grass or other undergrowth. These cosy microhabitats provide shelter from the wind, allowing the ambient temperature to increase above that of the surroundings. Sometimes, several snakes gather together in these pockets to bask. It may be that this group behaviour has multiple benefits, including safety in numbers or as a prelude to the mating season.

Shuttling

When snakes risk becoming too hot, they may seek cooler areas such as under leaf litter or logs, or in disused mammal burrows. Snakes often shuffle between warm and cooler habitats throughout the day in order to optimise their activities.

Snake diets

Britain's three native snake species are notably different in their appetites for prey, and this is probably no coincidence. Each species made its way to Britain after the last Ice Age receded 10,000 years ago, and it's likely that the success of all three since then has, in part, been down to the fact that there was relatively little overlap – and hence competition – in their tastes.

The amphibian specialist

If you see a snake hunting in water in Britain, it's almost certainly a Barred Grass Snake. Across its European range, this species specialises in hunting amphibians. As such, in Britain it hunts the Common Frog (*Rana temporaria*), Smooth Newt (*Lissotriton vulgaris*) and Palmate Newt (*Lissotriton helveticus*) as well as the recently introduced Pool Frog (*Pelophylax lessonae*). Interestingly, the Barred Grass Snake appears to have no problems swallowing poisonous amphibians such as the Common Toad, Natterjack Toad (*Epidalea calamita*) and Great Crested Newt (*Triturus cristatus*). It is likely in these

Below: A young Barred Grass Snake takes a frog tadpole. Amphibian larvae are likely to make up a large part of a juvenile snake's diet.

cases that both predator and prey have been embroiled in an evolutionary war of toxins that has lasted for many hundreds of thousands of years.

Undoubtedly, the Common Frog forms a particularly important part of the diet of the Barred Grass Snake. This is good news because frogs have adapted to become a regular garden species in many urban and suburban areas. It may be that, at least in some parts of Britain, the Barred Grass Snake might follow.

Although a specialist of amphibians, the Barred Grass Snake also regularly feeds on other wetland animals. Roach (*Rutilus rutilus*), Minnows (*Phoxinus phoxinus*), sticklebacks and ornamental goldfish are commonly hunted, as are nestling birds, particularly waterfowl.

The reptile specialist

The Smooth Snake thrives in habitats that support good populations of the Slow Worm, Common Lizard and Sand Lizard (*Lacerta agilis*). It also occasionally preys on other snakes, including the Barred Grass Snake and Adder, and is even known to take small mammals such as shrews and voles, and even fledgling birds.

Below: Having incapacitated a Slow Worm with its coils, this Smooth Snake begins the process of swallowing its prey.

Almost like a python, the Smooth Snake grabs its prey with its jaws, then wraps its coils around the animal to restrict its movement. Unlike true constrictors, however, the Smooth Snake does not kill its prey with this method. Instead, it often overpowers its prey and swallows it whole while it is still alive. Many of its kills are made in deep scrub or even underground in tunnels or burrows far from watchful human eyes. This secretive behaviour means that very little is known about the diets of young Smooth Snakes. It is thought that invertebrates form an important menu item in their early days of life, as is the case with our other native snake species.

The small mammal specialist

Unlike other British snakes, the Adder doesn't actively pursue prey, instead opting for more of a sit-and-wait strategy. Should a small animal accidentally stray too close, the Adder lunges forward with venomous jaws flung wide. Once struck, the prey item makes a dash, but within seconds the Adder's venom begins to take effect, leading to paralysis and, later, death. The Adder, meanwhile, slowly traces the scent trail of its fleeing prey, safe in the knowledge that its meal is secured.

Below: Though rodents are a vital food source, the Adder is also partial to amphibians and bird nestlings. Even weasels and moles have been known to be taken by this snake.

Although voles and other rodents make up an important part of the diet of Adders, they also take on amphibians and other reptile species, and hence their tastes are arguably broader than those of other British snakes. Adders have even been known to prey upon nestlings and eggs in the nest. However, the Field Vole (*Microtus agrestis*) is integral to its diet, along with the Wood Mouse (*Apodemus sylvaticus*).

Juvenile Adders feed on small amphibians, lizards and young mammals in their nesting burrows. Spiders and earthworms may also be taken, particularly by very young snakes.

Snake predators

Although snakes instil fear in many people, the truth is that the vast majority face a daily threat of being hunted and killed themselves – including by humans. British wildlife may be comparatively impoverished compared with mainland European fauna, but a number of snake predators are found here.

Above: Hedgehogs are one of a handful of mammals – including the Honey Badger (*Mellivora capensis*) and mongoose – that has evolved a degree of immunity to snakebites.

Hedgehogs

While the Hedgehog (*Erinaceus europaeus*) is adapted to consume invertebrates, it has been known to take on larger prey – including young snakes. Hedgehogs appear to have some immunity to Adder venom – while some bites are fatal, others appear to have no discernible effect. This suggests that the two species have overlapped within the same niches for hundreds of thousands of years.

Birds of prey

Undoubtedly, snakes are a food source for the Common Buzzard (*Buteo buteo*) and other large birds of prey. Interestingly, in many cases the raptors seem to be able to identify Adders specifically. Studies using plasticine models of the snakes to lure the attention of birds of prey suggest that the species' zigzag markings are a definite turn-off to hungry bird predators, which may have evolved to become wary of the Adder's defensive capabilities.

Right: A Common Buzzard carries a snake back to its nest. Many snakes carry scars from run-ins with the claws and beaks of birds of prey.

Pheasants

In addition to feeding on seeds and foliage, Pheasants are opportunistic hunters of small amphibians, reptiles and mammals. The impact of these birds – which are released in their thousands into the British countryside each year – on the reptile population has been debated for many years. In 2013, DNA analysis of Pheasant droppings at two sites in Worcestershire suggested that juvenile snakes may not be consumed in the numbers once feared. The debate, however, continues.

Cats

Questionnaire studies filled out by cat owners highlight just how effective the felines are at hunting urban, suburban and countryside populations of amphibians and reptiles, including snakes and lizards. In a single summer, British cats may collectively take more than 5 million prey items of these species.

Other mammalian predators

Other larger mammals that prey on snakes include Red Foxes and Badgers (*Meles meles*). In some cases, the Eurasian Otter (*Lutra lutra*) may seek out and eat grass snakes as they dip in and out of water.

Above: Juvenile snakes are especially vulnerable to predation by cats.

Below: The long, stabbing beak of a Grey Heron makes short work of all but the largest Barred Grass Snake.

Waterbirds

When spotted by water predators such as the Grey Heron (*Ardea cinerea*), the Barred Grass Snake often sinks to the bottom of the lake or pond, holding its breath among the weeds until the predator's interest wanes and it moves away.

Defensive duties

Venomous or not, British snakes use a variety of means to deter the interests of would-be predators. Some of these behaviours are common and some less so. With all of these strategies, however, there is a degree of overlap between the species.

Feigning death

When stressed by the unwanted attentions of predators such as cats, snakes sometimes opt to play dead, a behaviour called thanatosis. Although this may appear counterintuitive, it seems that removing the 'chase me' stimulus from a situation can cause some snake

predators to lose interest. Barred Grass Snakes are particularly drawn to this defence. Some individuals will lie on their back, almost with eyes rolled, letting their tongue fall out from their mouth theatrically. One 2007 study that looked at wild-caught Barred Grass Snakes showed that death-feigning behaviour was exhibited on 66 per cent of occasions.

Above: Attacks from cats often cause the Barred Grass Snake to feign death. If discovered like this, the best solution is to scare the cat away and leave the snake be. With the threat gone, it should quickly move off into the undergrowth.

Hissing

Snakes have a muscular ring (called the glottis) through which they control the flow of air in and out of the body. By constricting this ring and forcing air through it, they produce their characteristic hiss. All snakes have the physiology to hiss, but not all snakes employ a hiss when threatened. For some individuals, particularly smaller ones, it may be easier to make a dash for cover.

Spraying

Many snakes are able to release a strong-smelling fluid from their anal glands when threatened. The smell isn't wholly disgusting to the human senses, however – in fact,

I find there is a hint of garlic and mustard about it. Clearly, the defence works against animals with a stronger nose, such as dogs and cats. More than any other British species, the Barred Grass Snake appears particularly drawn to defending itself using this technique.

Striking

When threatened, some non-venomous snakes lash out with their jaws in a mock strike. This behaviour may be rare in Barred Grass Snakes, although there are reports of individuals biting people when poorly handled. The same is true of Smooth Snakes – *Reptiles and Amphibians of Europe*, written and published in 1962 by Walter Hellmich, describes this species as a 'vicious creature' that 'readily bites'. To avoid being bitten by a snake, venomous or not, the advice really is very simple: look but don't touch.

Above: Ready to strike! Many Adders do all they can to warn potential predators of their venomous intent before resorting to their defensive bite.

Fossil defences

There have been some reports of Barred Grass Snakes rearing up their head and flattening their ribs as if to make a threatening hood, like that of a cobra. This defensive behaviour is rather mysterious, since grass snakes (*Natrix* species) and cobras rarely overlap in their native ranges. Scientists have argued that, in the last 20 million years, the ranges of some of today's grass snake species once overlapped with the now extinct European Cobra (*Naja romani*). In other words, the mimicking behaviour has persisted in grass snakes, even though the venomous cobra they are imitating has long disappeared. Scientists describe this rare phenomenon as a fossil behaviour.

Parasites and fungal diseases

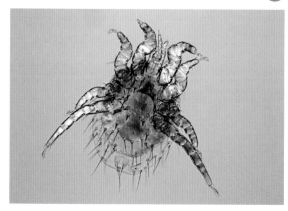

Above: *Ophionyssus natricis* is a common skin mite of reptiles.

Snakes are relatively secretive animals, and as such we still have a great deal to learn about their parasites. Internally, many snake parasites undoubtedly include the breeding stages of flatworms, roundworms and tapeworms, each picked up by swallowing amphibian hosts that act as carriers in the parasite's life cycle. Research into the parasite populations of wild snakes in other countries suggests that 75 per cent of adults may be infected in some way, either through gastrointestinal parasites, blood parasites or, perhaps quite commonly, tongue worms – worm-like parasites of a snake's respiratory tract.

All British snakes regularly suffer from parasitic mite infections on their external surfaces. These mites congregate around the only obvious cracks and crannies the reptiles possess – namely their eyes and the cloaca. In extreme cases, these infections can lead to serious bacterial infections and septicaemia.

Fungal diseases

Below: Scale damage caused by *Ophidiomyces ophiodiicola*, the pathogen which causes snake fungal disease.

In 2015, a fungal pathogen called *Ophidiomyces ophiodiicola* was discovered in the wild in Britain. Known previously only from North America, this causes a

condition popularly referred to as snake fungal disease (SFD) and in advanced cases leads to dehydration and lethargy. Close inspection of infected snakes may reveal that some scales on the underside of the body show signs of small brown lesions, each just 1–5mm (0.04–0.2in) in diameter.

Since the first discovery of the pathogen in the wild here in 2015, subsequent analysis of shed skins collected by volunteers has confirmed that it is present in many parts of England and is likely to be widespread. The impact SFD might be having on wild snakes is the subject of important research currently being undertaken by the Garden Wildlife Health initiative (see page 126).

Fungal infections like SFD can undoubtedly travel between wild populations and captive snake populations, so it is important that reptile owners maintain good biosecurity practices.

Battle scars

Snakes have an impressive knack for recovery. Although they take longer to heal than mammals due to their slower metabolism, they can recover from some apparently serious encounters with predators. Their injuries can include broken bones, skin tears from claws or bite marks from teeth, and occasionally they end up with a stump at the end of their body where their tail has been bitten clean off. Sometimes, individuals apparently close to death – for instance, with deep gouges left from an attack by a bird of prey or an accident with a garden strimmer – can live on for months or even years. These 'battle scars' remain as the snake grows, because new scale tissue forms over the top. Such scars can be a handy way to identify individual snakes in your patch.

Above: Making a note of battle scars can help for quick identification of individuals within a population.

Snakes and the Seasons

In the UK, the cold winter months offer a double blow for reptiles. First, because they are unable to generate their own body heat, their physiology slows. And second, most of their prey either dies off or disappears underground. It makes sense, therefore, for snakes to sleep through winter, but doing so presents its own problems. It means snakes must achieve in six months what many mammals and birds typically manage to do in 12. So, from their first emergence from hibernation, the pressure is on.

Emergence

Although the term hibernation is commonly applied to describe the behaviours of dormant reptiles in winter, the intricacies involved separate them from overwintering mammals, the true hibernators. Instead, reptiles are better described as brumators.

First, let's look at the similarities. Both brumation and hibernation begin because of shortened day length and increasingly cold temperatures. Likewise, both behaviours involve the reduction of heart rate and the lowering of body metabolism. And finally, both behaviours cause a period of starvation that may last many days or weeks. The edge is taken off these periods of zero-energy intake through the use of fats in the body, which act as a temporary energy source.

Yet brumation does differ from true hibernation. For a start, snakes can use glycogen as an additional energy source while they slumber. These sugars are stored in their muscles and provide for another stroke of fortune. Through a fascinating quirk of biochemistry, the presence of glycogen in reptile blood reduces the oxygen

Above: Adders can be very sluggish when first emerging after winter. Keep your dogs on a lead on sites in which Adders are known.

Opposite: A large Barred Grass Snake gathers much-needed warmth after emerging from a nearby disused mammal burrow where it has spent the winter months.

Above: As well as collecting radiation from the sun, this Smooth Snake is likely collecting heat being radiated from the sun-drenched log upon which it rests.

requirements of these animals. This means that snakes can hibernate in lower oxygen environments than mammals.

For British reptiles, overwintering sites are always in locations where frost cannot penetrate. These sites may include Red Fox, Rabbit (*Oryctolagus cuniculus*) and rodent burrows or the cracks among tree roots. Some reptiles favour human-made constructions that mimic these natural conditions. The ruins of old farm buildings, for instance, offer plenty of cracks and crannies in which they can hide. In addition, piles of rocks, straw, manure or old building materials may occasionally be used. Sites that are sheltered from the wind and away from spots prone to flooding may end up being used for generations by numerous species of reptile, not just snakes.

Most British snakes make their first forays out of their winter dens on sunny days between March and April, although emergence can be earlier in southern and western counties. Males of all British species use the early spring sun to help bring on their sperm, and for this reason they are often spotted by human observers ahead of females.

Matters of mating

As April turns to May and the weather warms up, the snakes' interests soon turn to courtship. This is an important time for snakes, particularly in temperate regions like Britain, where females may choose to mate only once every two years or so. For this reason, competition among males for mates really does heat things up.

Although it's easy to consider snakes as largely solitary or skulking animals, it's clear that at some times of the year they keep track of one another very closely. They do this mostly through smell. The primary mode of communication for snakes is through special odour molecules called pheromones, which are deposited as they move through the undergrowth. During the breeding season, pheromones may indicate the species, size and sex of the individual, providing a handy trail for potential mates to follow. Snake skin also increases in importance at this time as a provider of yet more pheromones. In fact, the females of some species shed their skin especially early at this time of year so that the scents produced are as fresh as possible.

Below: A male (left) and female (right) Adder enter the final stages of copulation. At this point, sexual proceedings are largely governed by sensations of touch and smell.

Above: During the mating season, Barred Grass Snakes become so preoccupied with proceedings that they are easy to observe with low-powered binoculars.

Pheromone trails don't linger indefinitely, however – as several males home in on a female, for instance, their intensity may weaken. For this reason, male snakes also depend on visual cues to find females and can be observed carefully scanning their surroundings at regular intervals. Scientists call this early part of the snakes' reproductive behaviour the pre-courtship period. Courtship properly begins when a male and female come face-to-face for the first time.

During courtship, the male faces the female head on and flicks his tongue with a rapid intensity not seen at other times of the year. If the female doesn't move away, the male makes physical contact, moving his chin gently along the top of her body. At this point, males of many snake species appear to produce special waves of muscular movement (called caudocephalic waves) that ripple up or down the body. These contractions may be important in signalling the stage to come: mating.

As with all snakes, fertilisation in the three British species is internal. Males are armed with a special pair of copulatory organs, called hemipenes, which can be everted from the cloaca and through which sperm travel into the female's reproductive tract. Interestingly, females of many snake species are capable of storing sperm from previous matings, allowing them to refertilise a new batch of eggs months or even years after a mating occurs. How often this tactic is used in our native species is not yet known.

During mating, male and female snakes may be entwined together for an hour or more. Eventually, when conditions allow, they uncouple and move off in separate directions, almost as if no union ever occurred.

Dancing for delights

Of course, a female may not choose to mate with the first male she comes across. For her, it can pay to delay proceedings in case a stronger male (with a better hand of genetic cards, so to speak) should come along. Should this occur, a battle of wills and strength must take place between males, after which the winner might be afforded the gift of genetic posterity.

Male Barred Grass Snakes and male Smooth Snakes regularly engage in bouts of hissing, posturing and tussling between one another, but the Adder displays the most enduring and athletic of sexual battles at this time of year. Male Adders regularly rear up against their opponent, squaring up to the other's eyeline, before attempting to push the rival to the ground in an act of unrivalled strength. While undertaking this strange behaviour, the warring snakes regularly coil around each other, wrestling, flipping and tumbling together like Hollywood stunt actors over-egging a barroom brawl. Predictably, the athletic endeavours of pre-copulatory Adders have taken on almost mythical significance among nature lovers, who whimsically refer to the event as the 'Dance of the Adders'.

Below: Competition between males to find females can reach fever pitch in spring. Here two male Adders attempt to show their strength by pushing one another to the floor with their heads.

Sexing snakes

Above: A female (left) and male (right) Adder. Note the more contrasting colours in the male. Females are, on the whole, varying shades of bronze-brown.

Of the British snake species, male and female Adders are by far the easiest to tell apart. Simply put, male Adders have black stripes on a silvery-white background colour and females have dark brown stripes on a bronze background. Sexing the Barred Grass Snake and Smooth

Right: In common with many snakes, the largest Barred Grass Snakes are likely to be female.

Snake is far more difficult. Indeed, even experts sometimes get this wrong.

In males of all species, a bulge near the cloaca hints at the presence of the hemipenes tucked away inside the body. In addition, the tails of male snakes are said to be slightly more elongated and tapering than those of females. Large female Barred Grass Snakes are said to lose the yellow coloration of their distinctive collars, although the black edging remains.

Right: As with all snakes and lizards, the reproductive anatomy of snakes is paired – these are the so-called 'hemipenes'.

Deferring investments

Living in a temperate climate comes at an obvious cost to British snakes. Most notably, the lack of feeding opportunities and the cooler conditions mean that females of all British species regularly skip a breeding year in order to sustain their breeding weight. While female Adders in southern Europe, for instance, may breed annually, those in Britain often breed only once every two years. Interestingly, in the Alps, where temperatures are cooler owing to the high altitudes, female Adders may breed just once every three years. Although this behavioural adaptation allows snake species to colonise and thrive in areas where other reptiles may struggle, it can leave them at greater risk of local extinction. Put simply, species that breed infrequently struggle to get their numbers up after bad years, dampening the long-term survival chances of a population.

Above: In Britain, female snakes regularly forego breeding seasons to maintain a healthy body weight.

Sliding into summer

Right: Compared to other British snakes, the Smooth Snake is a slow coloniser of new haunts. This makes this species more dependent on help from conservationists.

Below: Adders often move through their habitats using defined corridors, such as sun-drenched banks and woodland rides.

At many sites across Britain, the movements of snakes change during the summer months when the consistently high temperatures offer more predictability by way of hunting opportunities. Adders, for instance, may move into wetter areas such as valleys or bogs, and Barred Grass Snakes may move between freshwater sites, seeking out untouched spoils. At this time of year, both of these snake species are capable of travelling several hundred metres in a single day, and their home ranges occasionally take in 50–100ha (125–250 acres). Smooth Snakes, on the other hand, don't travel quite as far. Each individual may roam just 10–20m (11–22yd) beyond its regular patch, and that's about it.

Aestivation

As discussed earlier (see pages 57–8), snakes operate within a fine window of temperatures. Clearly, cold weather has an impact on their activity, but the same is also true of hot spells. During drought years, when rainfall is low and temperatures are consistently high, snakes become prone to dehydration and many enter a period of enforced inactivity called aestivation. To all intents and purposes, this is a bit like a summer hibernation period. To reduce water loss, aestivation normally occurs in places where humidity remains high throughout the day, such as tree burrows in valley mires or shady woodlands.

Birthing plans

For the most part, British snakes give birth to their offspring in the summer months. Various branches of the reptile family tree have evolved to give birth to live young rather than laying hard-shelled eggs, a behaviour that, at first glance, appears almost mammal-like. At birth, these young snakes are enclosed in a thin, opaque membrane called the pellicle, which readily splits apart to release the snake into the world. Although this method of delivery resembles mammalian birth, these reptile foetuses lack a placenta and instead receive nourishments from a yolk sac. In zoological terms, reptiles that produce offspring in this way are called ovoviviparous.

Of the British snakes, two species – the Adder and the Smooth Snake – are viviparous and give birth to live young. The Adder gives birth sometimes as early as July or as late as early October to anything from three to 20 offspring, each one averaging the length of a pencil (16cm/6in). The Smooth Snake gives birth in August or September to between four and 15 offspring that are of similar size to Adder newborns.

Below: An Adder gives birth to its young. Unlike in mammals, within moments of birth, a young snake is ready for action.

Above: A young Adder breaks free from its pellicle. Baby snakes of all British species resemble the adults in colour and markings. Each hatchling is roughly pencil-like in size and stature.

In vertebrates, ovoviviparity has evolved in many separate species (almost 150 in total) so this method of delivery must have distinct evolutionary advantages. In the Adder, ovoviviparity undoubtedly helps keep eggs warm and at a stable temperature, allowing the species to colonise cooler environments than other snakes.

Right: Young Smooth Snakes make their appearance in September. Comparatively little is known about the daily behaviours of young snakes.

Egg activity

The Barred Grass Snake is the only native snake to lay eggs with a hard shell, producing clutches of up to 40 or so. It lays them in rotting piles of vegetation that generate plenty of warmth for incubation. Natural nesting sites are in vegetation on the sun-drenched strandlines of rivers and lakes, but manure piles and garden compost heaps have been added as new (and very valuable) options for egg-laying females. Occasionally, however, Barred Grass Snakes find other locations in which to lay their eggs, including mammal burrows, the foundations of walls and old buildings, and even (on more than one occasion I know of) school sandpits.

Above: A female Barred Grass Snake with her clutch of eggs. Large individuals can lay up to 40 eggs in a single sitting.

Left: A sunny compost heap with lots of access points – a classic garden egg-laying site for the Barred Grass Snake.

Each egg is 2.5cm (1in) or so in size and is cemented to the others around it to add a degree of protection from the elements. Often, several females will use the same egg-laying site, leading to the sudden emergence of hundreds of tiny hatchlings 10 or so weeks later. Each foetus has a tiny egg tooth – a tough spine on its nose that helps it to split open the shell. Leftover eggshells often remain stuck to one another, taking on a texture a bit like miniature crushed-up ping-pong balls. Occasionally, these spent eggshells are discovered by gardeners turning over their compost heap in autumn.

The survival of an entire clutch of Barred Grass Snakes depends on the female finding a suitable egg-laying site that remains both warm and predator-free. For this reason, she will spend much time investigating potential locations for her offspring in early summer. Even with this due diligence, there are some things these snakes can't control – notably the great British summer weather. Undoubtedly, the cold and rainy summers we occasionally have can be a real problem for breeding snakes locally. As with many reptiles across the UK, there are good years and bad years for our native snakes.

Below: The natural warmth generated by trillions of hungry microorganisms makes manure heaps another favourite egg-laying site for the Barred Grass Snake.

Left: A young Barred Grass Snake sets its eyes upon the world for the first time – only minutes old, it is already watchful for potential predators.

Below: Once dried-up, spent Barred Grass Snake egg shells feel almost like plastic. Occasionally, some contain embryos that have failed to survive to hatching.

Snakes on the Slide

The fact that only a tiny proportion of people living in Britain have ever seen a wild snake tells you all you need to know about the reptiles' secretive way of life. But it also hints at the declining fortunes of snakes in recent years. Through historical accounts and records, it's clear that snakes used to be seen far more frequently by members of the public across many parts of Britain. Something has happened to them in recent decades – but what?

A global problem

Like many reptiles, snakes are in big trouble. Internationally, of more than 3,000 named snake species, approximately 200 or so are threatened in some way with extinction. In addition, many hundreds more are likely to be declining through overharvesting for traditional medicines or the illegal pet trade, through habitat loss or through the introduction of invasive predators to their habitats.

It's worth highlighting that the threats to our British species are notably different from those affecting snakes in other parts of the world. Because Britain is a relatively small island with moderately limited space, our snakes arguably rub up against humans and our complex infrastructure more frequently than those living elsewhere. In many cases, seemingly small landscape changes have made life impossible for local populations of snakes, and they have died off without anyone really even noticing. But notice we must. For the more we observe their populations and protect their habitats, the better their survival prospects are likely to be. All is not lost – not yet, anyway.

Above: The Antiguan Racer (*Alsophis antiguae*) – one of 200 or so snakes threatened with extinction. This number is likely to grow in coming years.

Opposite: More than ever before, snakes across the world must find a way to coexist with an ever-increasing human population.

The fate of British snakes

Right: A car park represents dead space to local snakes. At best, it is an inconvenience; at worst, it is a risk to snakes and a barrier between populations, limiting gene-flow and making local extinction more likely.

A handful of statistics is all it takes to reveal the losses British snakes have endured in recent decades. The first statistic is that, in a single century, we have lost almost three-quarters of our wild ponds, mostly through drainage and infilling. For amphibian hunters like the Barred Grass Snake, this has undoubtedly hit hard. The second statistic? Across the Thames basin, parts of Sussex and throughout Dorset, we have lost almost 90 per cent of our lowland heathlands, a traditional stronghold for both the Adder and Smooth Snake. The third statistic is perhaps even more alarming because it refers to very recent changes that have been occurring year on year as our urban areas expand. In the six years between 2006 and 2012, for instance, 2,250km² (870 square miles) of forests and farmland were cleared to make way for artificial (urban) surfaces such as shopping developments, housing and industry – a habitat snakes have obvious problems making their own. These land-use changes are continuing right now as you read these words. Our snakes, being poor dispersers and having quite specialist needs, are unlikely to be adjusting well.

At this point, you may be thankful that we have nature reserves and other rural spots throughout the UK dedicated to preserving wildlife. But these sites

Below: Snakes can become exposed to predators – particularly birds of prey – when moving across roads and paths.

Above: It may be that, in future, wildlife tunnels or 'green bridges' help link up populations of reptiles separated by roads such as this one.

can come with their own problems for snakes. Many heathlands, for instance, are naturally inclined to move towards the next level of succession, becoming shady forest-like habitats less suited to snakes. Roads that bisect such sites cause reptile populations to bunch up, resulting in genetic stagnation and inbreeding. Even nature-loving humans cause problems here, bringing with them over-inquisitive pets that disturb snakes or, worse, prey on them.

Thankfully, snakes have some protection by way of UK law. Each of our native British species is protected under the Wildlife and Countryside Act 1981, which makes it an offence to kill or injure them intentionally. Although this decades-old piece of legislation has proved helpful to our snakes, the key word here is *intentionally*. Sadly, the unintentional killing of reptiles is likely to have continued apace over the last 40 years, particularly through the poor management of many important reptile sites.

More protection for the Smooth Snake

Being rare and far more isolated in its natural habitats than our other British snakes, the Smooth Snake currently receives a higher level of legal protection in UK law. As well as being an offence to capture, possess, trade or kill or injure a wild Smooth Snake, it is also illegal to damage or destroy Smooth Snake breeding or feeding sites. Although the species is widespread across many warmer parts of Europe and even into Asia, this high level of protection is required to secure a future for the Smooth Snake on our islands.

Death by a thousand cuts

Ideal reptile habitats include a mosaic of microhabitats, within which there are places to bask and places to keep cool, as well as places with abundant prey, lots of places to hide and places to hibernate. Provided these features persist on a given site, snake populations should survive – but managing this can be a precarious balancing act. Should any of these habitat features fail or be removed accidentally, the local snake populations will be set on a downward spiral towards extinction. This is likely to have occurred across many wildlife sites in recent decades – bit by bit, snakes across England, Scotland and Wales have died a death by a thousand cuts. The combined threats facing Britain's snakes are explored below.

Climate crisis

Reptiles such as snakes are likely to be very sensitive to the changes that the climate crisis might bring. Much will depend on how the British climate changes, which is something we can't yet predict with much degree of accuracy. If we have longer, drier summers – as some models predict – it's possible that the rate of survival of snake offspring may increase, particularly in the case of

Below: Climate change puts extra pressure on British reptiles, making their successful conservation more uncertain than ever.

Above: A cold winter may be a good thing for over-wintering reptiles – colder months slow their metabolism, saving precious energy which can be used in springtime.

the Barred Grass Snake. However, drought brings with it its own problems – most notably fires (both intentional and accidental), which already regularly rage across Britain's heathlands each summer. Likewise, the climate crisis may also affect our winters. Milder temperatures at this time of year are not always good news for snakes because a warm snake that is brumating underground burns up more energy than a cold one doing the same thing. This could lead to lower breeding weights for wild snakes come spring, a phenomenon already seen in Common Toads living wild in Britain.

Traffic

Although snake mortalities on our roads don't receive as much attention as the deaths of other wildlife such as Hedgehogs or toads, it can be an everyday occurrence at some sites, particularly where a roadside bank acts as a favoured basking location. At one site in Poland, 190 dead Grass Snakes were recorded in a single 1.8km (1.1-mile) stretch of road over a period of just 10 months, and 89 per cent of them were juveniles.

Below: It is possible that some people actively drive over snakes when they see them on roads, even though this is technically illegal under the Wildlife and Countryside Act 1981.

Above: Understanding more about the genetics of wild populations means that scientists can assist and inform conservation efforts for threatened snake species in Britain.

Below: A snake having its DNA sampled. The procedure involves scientists taking a swab of cells from the snake's vent that is taken back to the lab for analysis.

Genetic stagnation

In Britain, many snake populations are split into discrete pockets of land that are hemmed in on all sides by roads and other human infrastructure. If a given snake population is small, this isolation can lead to inbreeding over time, a situation that can reduce the resistance of individuals to diseases and the overall survival of offspring. In recent years, scientists have been interested in how much of a problem so-called inbreeding depression might be for our snakes. It is thought that the Adder, in particular, might be prone to this threat, especially given that a third of remaining Adder populations are thought to comprise fewer than 10 adults. Most notably, Adder populations in the English Midlands are particularly fragmented. Continued research into these populations – involving scientists undertaking DNA swabbing – is likely to shed light on how serious genetic inbreeding may turn out to be for our snakes.

Disturbance

At some sites, dogs running off the lead at certain times of year, for instance, may disturb individual reptiles as they bask. On some nature reserves where Adders are common, this can lead to an increased risk of pets being bitten. Sometimes, even the interests of people (some armed with camera phones and eager for the perfect shot) may cause snakes periods of heightened stress.

Above: Many important reptile sites have signs that inform visitors to keep their dogs on leads.

Introduced predators

Love them or loathe them, cats are sublime predators of native species (see page 65). According to one five-month survey undertaken in 1997, it's possible that they may collectively kill as many as one million reptiles each year in Britain, a proportion of which will be snakes. On some important reptile sites, cat-proof fencing may help snakes survive in peace, but the costs of maintaining such fences can prove prohibitive for some landowners.

Below: Though reptiles can be fast and agile in the summer months, some will fall prey to introduced predators such as cats.

Persecution

My well-worn copy of the 1921 Edward Step's classic *Animal Life of the British Isles* has within it an excerpt from a troubling news story that reads as follows:

> An enormous snake was killed yesterday ... only a few yards from where some children were playing. The Rev. Mr. Blank courageously seized the reptile behind the head, but when it hissed savagely at him he was forced to throw it down. Its head was then smashed with a pole, and finally it was dispatched with the aid of a spade. The venomous monster was found to be over three feet [1m] in length. Its nest was found and a large number of eggs destroyed.

Below: A dead adder hanging from a tree, an alleged victim of snake persecution in East Yorkshire.

The story is a distressing indictment of a once classic, very British (and now illegal) response to snakes. More troubling still is that the species in question was undoubtedly not a venomous Adder but a harmless Barred Grass Snake – the only egg-laying snake native to Britain and the only species that can grow to 90cm (35in) in length. Reading this excerpt today, it might be tempting to assume that our public response to snakes has changed, but cases of alleged snake persecution are reported to the police in the modern day, albeit very rarely. We have work to do yet, to inform the public of the protected status of reptiles in the UK and their legal right to a life without risk of intentional harm and killing.

Thankfully, wildlife organisations and conservation volunteers are playing an increasingly important role in keeping scare stories about wild snakes grounded and based on truth. The key messages the media need to relay are that British snakes are declining and have as much

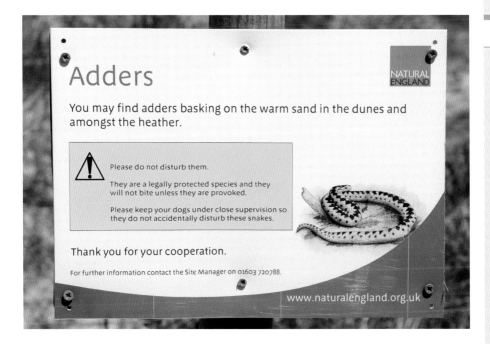

Adders

NATURAL ENGLAND

You may find adders basking on the warm sand in the dunes and amongst the heather.

⚠️ Please do not disturb them.

They are a legally protected species and they will not bite unless they are provoked.

Please keep your dogs under close supervision so they do not accidentally disturb these snakes.

Thank you for your cooperation.

For further information contact the Site Manager on 01603 720788.

www.naturalengland.org.uk

right to be here as anything else, that all British snake species actively avoid humans and their pets if given the chance and that our snakes are protected by law from being injured or killed.

Increasingly, we are making strides to limit snake persecution in Britain, but there is still much to do. Snakes have a surprising amount of reputational baggage that needs to be shaken off before they can be truly taken into the nation's heart, so continue we must.

Above: Disturbance is a big issue to reptiles. Moving from place to place burns up valuable fat reserves and makes them susceptible to being spotted by predators, including birds of prey.

Fire

Uncontrolled accidental fires can be a significant threat to heathland sites that support rare reptiles, including species such as the Smooth Snake and the Adder. Sometimes these fires can start through simple accidents, such as from a badly tended barbecue or a thoughtlessly discarded lit cigarette. On other occasions, sadly, fires are started deliberately. The reason that unplanned fires like these are so serious is because, in extreme cases, they rage out of control, destroying great swathes of important reptile habitats and leading to the local extinction of

Above: Increasingly, uncontrolled heathland fires are receiving the media attention they deserve. Such fires may become more likely as climate change influences the frequency of drought in the UK.

snakes and other rare species. This isn't to say that all fires on heathlands are bad for reptiles, however. Some conservation organisations, including the RSPB, use small-scale controlled burning as a management tool to refresh small patches of heathland in which there are no known reptile hibernacula. These organisations undertake heathland management practices like this with great care and planning.

Right: As well as directly killing reptiles, wild heathland fires turn a complex ecosystem into a featureless wasteland. This is not a good place for reptiles to hunt or to hide within.

Non-native diseases

In recent years, the transmission of wildlife diseases from continent to continent has been implicated in the decline of a number of species. A fungal parasite that causes an amphibian disease known as chytridiomycosis has ravaged thousands of frog, toad and salamander species worldwide, for instance, and bats have been hit hard by another fungal pathogen that causes a disease commonly known as white-nose syndrome.

Snakes, it seems, can also succumb to fungal disorders. Indeed, snake fungal disease (SFD), first found in North America, was detected for the first time in Britain's wild snakes in 2015 (see page 68). As discussed earlier, SFD is caused by the fungus *Ophidiomyces ophiodiicola* and leads to a number of symptoms, including lesions and crusty scales on the skin of infected snakes. In some cases, infection with SFD may reduce general overall fitness of the snake and, in severe cases, even lead to its death.

Below: By moving snakes between different countries in less-than-adequate conditions the exotic pet trade is likely to be a significant route through which wildlife diseases might find their way into new countries.

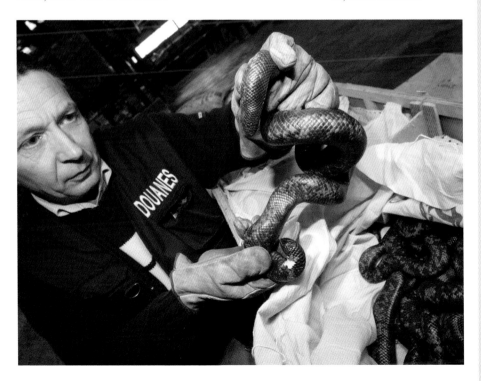

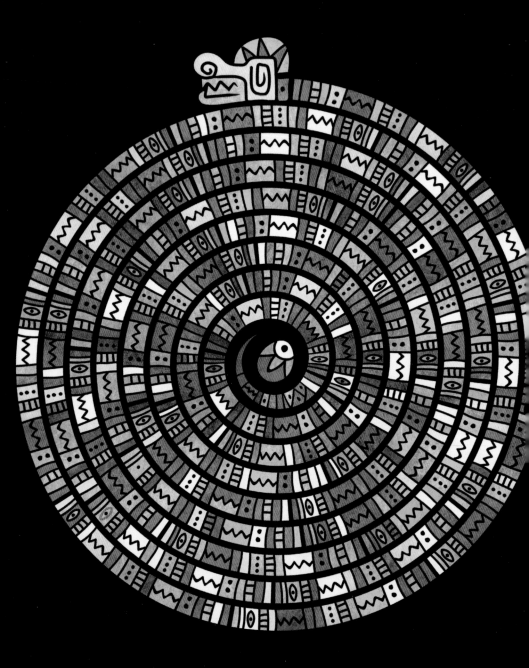

Snakes in Culture

It doesn't take much historical investigation to confirm humankind's fascination with snakes – they repeatedly feature in our myths and legends. According to many of these legends, snakes are untrustworthy creatures cursed by dark magic or are even considered to be evil incarnate. But if you're willing to delve deeper into our history, it becomes clear that not all humans hold snakes in such hateful regard. And at one time in our history, snakes were objects of fertility or the stuff of creation myths – of life everlasting.

Sacred snakes

If there is one unifying feature of the human species, it is our interest in danger. A potential threat – be it a strange van in the street or a wasp in the house – sparks interest in us. Clearly, snakes (venomous or not) have always elicited a similar reaction, but they also tick another box: they are mysterious and hard to see. Snakes seemingly appear out of nowhere, almost as if guided by magic. And, with their reflective scales and forked tongue, they are novel. They stand out in stark contrast to the mammals and birds that surround us. For these and other reasons, snakes feature prominently in the world's myths and stories.

One of the most frequent serpentine creation myths, particularly in Africa and Australia, is that of the Rainbow Snake, a colourful entity responsible for birthing the world's animals. According to some of these myths, the figure is a giant water god whose slithering movements carved out the world's rivers and valleys. Interestingly, this pervasive story has parallels with a Greek cosmological myth in which Ophion, a serpent created by a supreme goddess, incubates the primordial egg from which all life subsequently comes. These are not the only snake-related creation stories, either. In Chinese mythology, for instance, Nüwa, a snake with

Above: Snakes feature heavily in Aboriginal cave art. In fact, these works may represent some of the oldest examples of Stone Age art undertaken by modern-day humans.

Opposite: Snakes abound in many of the world's creation myths. This might be because of their otherworldly appearance or, perhaps, because their elongated appearance lends itself well to the notion or concept of time.

Above: Artworks produced by the indigenous peoples of Australia regularly feature snakes, often in vibrant colours.

the head of a woman, was responsible for creating all of humankind, each person being a lump of clay breathed into life. According to the myth, Nüwa hits upon a shortcut in the production process. By dipping a rope in clay and flicking it, the land became populated with tiny

Right: An example of Aboriginal artwork featuring the Rainbow Serpent from Kakadu National Park in Australia's Northern Territory.

blobs, each of which became a person. The first time Nüwa flicked her rope of clay, the humans produced were high class. Predictably, the second time she flicked her rope, the blobs of clay that splattered across the ground gave rise to the lower classes.

Like frogs, snakes often feature as symbols of fertility. For instance, the Hopi people of North America collect snakes and handle them in a special dance that celebrates the union of a sky spirit called Snake Youth and an underworld spirit, Snake Girl. At the end of the ceremony, the snakes are released into the fields to help secure an abundant harvest. This sort of reverence for snakes also occurs in other parts of the world. In east Asia, snake-dragons watch over crops, while in ancient Greece and India, snakes were considered talismans that provided protection against evil – the opposite of how many perceive the reptiles today. How snakes became associated with fertility and seasonality may never be known. Some argue it is because they can form a circle – a symbol of the cycling seasons. Others suggest it is because they shed their skin, in what can be considered a metaphor for life renewing itself.

Below: A more modern interpretation of snakes from the Museum of Old and New Art in Tasmania, Australia. This artwork by Sir Sidney Nolan features 1,620 panels and takes its inspiration from the desert in springtime.

Upon thy belly

The LORD God said to the serpent, 'Because you have done this, cursed are you above all cattle, and above all wild animals; upon your belly you shall go, and dust you shall eat all the days of your life. (Genesis 3:14)

Although it's tempting to lay all of the blame for our apparent hatred of snakes upon this classic biblical verse, many other ancient stories paint snakes in a less than favourable light. In some of these cultural stories, particularly from India, snakes were associated with drought. In one example, the drought serpent Vritra (or Ahi) swallows the primordial ocean. It is then killed by the warrior Indra, thereby releasing the water (and the animals contained within it) across the land. In another story from India, the world's earthquakes are blamed on the yawning and stirring of a giant world serpent called Shesha.

Snakes often feature as the guardians of mystical underworlds in myths and legends. In Greek mythology, these guardians took the form of the Gorgons, three sisters (including Medusa) who had live venomous snakes for hair and whose gaze could turn those looking at it to stone. In Hinduism, Buddhism and Jainism, the naga and nagini were a semi-divine race of human-snake hybrids that lived in a heavenly netherworld called Patala.

Below: One of many artistic representations of that fateful day in the Garden of Eden.

In these Greek and Indian myths, snakes – though pitched as unpredictable and mysterious – are empowered, respected animals that are given a certain agency in their actions. This contrasts sharply with the Biblical view of snakes as little more than cursed, cunning and untrustworthy beasts. Sadly, snakes in the Western world have had to carry this Biblical baggage with them ever since.

To infinity and beyond...

Scholars have long debated the interpretations of snakes in mythological stories, and none more so than Ouroboros, the famous symbol of a snake eating its own tail. This icon first turned up in Egyptian funerary text in the 14th century. Still, it appears to have spread throughout a range of cultures and has stacked up a dizzying array of interpretations. In late Egyptian times, Ouroboros came to represent the cyclical nature of the year. To the Romans, it indicated magic and fortune. And to the Gnostics, it performed a role analogous to the Taoist yin and yang of existence.

The Ouroboros was even used as a symbol for the philosopher's stone, a legendary substance sought by early alchemists. In Norse mythology, the Ouroboros was depicted as a serpent born of a god, which grew so large that it encircled the world and grasped its own tail in its mouth. Interestingly, a similar tale is told by indigenous people in parts of the lowlands of South America. To those cultures, the world was a disc whose circumference was straddled by a giant anaconda that, once again, was thought to hold its own tail in its mouth.

Above: Ouroboros, the snake that eats its own tail.

A snake in the grass

You can tell a lot about the medieval perception of snakes by looking at how the Old English word 'snake' came to become more than just a noun. As the centuries of the Middle Ages passed, use of the word 'snake' (or worse, 'snake in the grass') became more and more common to describe people capable of deception. Records show that, by the late 16th century, the use of 'snake' in this way was commonplace. Snakes had become, to all intents and purposes, a bad thing.

However, snakes did briefly find a more positive role in society at this time, depending on your perspective. Like frogs and toads, to which they were once assumed to be closely related, snakes (particularly Adders) periodically found their way into questionable folk

Right: A 13th-century Arabian artwork featuring a man bitten by a snake and surrounded by plants believed at the time to be a cure for venom.

remedies. One such therapy involved dipping a bag of snake, toad and newt heads in water and then dripping the resulting fluid onto the areas of the body requiring treatment. Thankfully, in Britain at least, this newfound role as a potential medicinal saviour wasn't to last.

What happened in the centuries that followed should come as no surprise. Enter the infamous snake-bashers of the Victorian era. These individuals were paid to catch wild snakes, and most operated in the New Forest. The most famous of this motley crew was Harry 'Brusher' Mills (1840–1905). At some point around 1880, Brusher relocated to a small hut in the middle of the New Forest and began a new snake-catching enterprise. In the two decades that followed, he is thought to have caught more than 20,000 snakes, pinning them down with a trusty forked stick, before picking them up and dropping them into a sack.

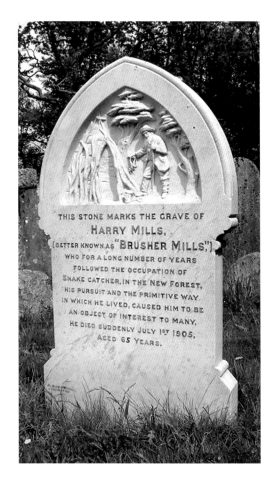

He sold many of these snakes to London Zoo, where they were fed to various snake-eating snakes and birds, or he peddled them to purveyors of Adder-fat, a substance once used as a treatment for snakebites.

Above: A marble headstone erected at the grave of 'Brusher Mills' in the churchyard of St Nicholas in Brockenhurst. Brusher is likely to have killed as many as 20,000 snakes.

Some argue that Brusher must have developed an immunity to snakebites during this time. We will likely never know, but the copious amounts of rum he consumed was said to have played a role in his recovery (a local pub in which Brusher regularly drank was renamed The Snake Catcher in his honour). After Brusher's death, two more generations of snake-catchers persisted in the New Forest, each using the familiar tools of the trade: a simple forked stick and a sack.

Snakesploitation

Nothing catches the eye quite like a potentially dangerous snake. In fact, as a group, snakes are capable of enchanting and terrifying us in equal measure. For this reason, the reptiles are sometimes used as a means of attracting immediate interest from the public, as a prelude to financial profit. Generations of street performers, questionable medicinal practitioners and even celebrities have benefitted from exploiting snakes in this way.

Snake charming

Once performed throughout India, the roadside snake-charming show involves a 'charmer' playing a flute-like instrument that apparently brings a swaying snake out of its slumber from within a basket. The movement of the musical instrument is actually what the snake responds to, not the tune (which it cannot hear). The practice is thought to be dying out as laws have become more strict and other forms of street entertainment have taken over.

Below: Snake-charmers working their trade. Though the practice has cultural significance, modern-day snake charmers are rarer than they once were.

Snake cocktails

In many parts of the world, snake blood (particularly cobra blood) is a component of alcoholic cocktails aimed at increasing sexual virility. In these cases, the blood is drained from live snakes.

Snake consumption

Though most cultures avoid eating snake meat, in some parts of the world it is considered a prized local delicacy. Cooked rattlesnake meat, for instance, is consumed in the Midwestern United States and snake soup is a Cantonese delicacy especially popular in Hong Kong during the winter months.

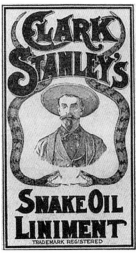

Above: Once marketed as an all-curing tonic, nowadays the term 'snake oil' is a euphemism for deceptive marketing.

Left: Snake soup is made from snake meat and bones, simmered for many hours. Historically, the dish was used to warm up the body on cold days.

Newspapers

Nothing sells a story quite like a monster on the loose, and no media domain knows this better than the newspaper industry. 'Adder Attack!' and other sensationalist headlines on a front page can more than double sales, so it's in the interest of many tabloids to add a healthy dollop of drama to each and every incident involving snakes. Critics argue that headlines like these create a negative feedback loop in the minds of the public – each 'Shock horror!' encounter breeds more fear of snakes, which drives more newspaper sales the next time such an event occurs.

Serpentine cinema

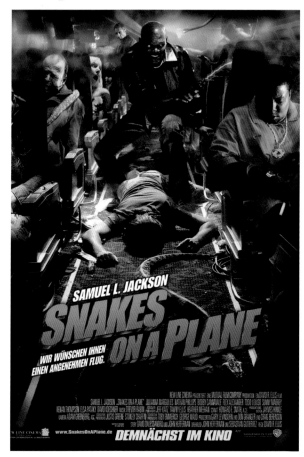

No storytelling medium has painted snakes in such a negative light as Hollywood cinema. Since the rise of the movie industry more than a century ago, snakes have become the go-to animal for cheap scares and bargain-bucket horror.

During this time, snakes have been shot in *American Sniper* (2014), *Bad Girls* (1994) and *Fools Rush In* (1997), eaten in *Capricorn One* (1977) and *Indiana Jones and the Temple of Doom* (1984), impaled in *Anaconda* (1997), possessed in *Beetlejuice* (1988), swallowed for punishment in *Collateral Damage* (2002) and used for curses in *Fantastic Beasts: The Crimes of Grindelwald* (2018); see also *The Curse II: The Bite* (1989), *The Reptile* (1966) and *Snake Woman's Curse* (1968). They have also been hacked to death in *Friday the 13th* (1980), used as props for exotic dancers in *Bladerunner* (1982) and *From Dusk Till Dawn* (1996), and masqueraded as makeshift ropes in *Indiana Jones and the Kingdom of the Crystal Skull* (2008). Snakes have been sliced and diced in *Prince of Persia: The Sands of Time* (2010), totally incinerated in *Superman II* (1980) and, on more than one occasion, sucked out of planes in *Boa* (2001) and *Snakes on a Plane* (2006). The list could go on – and that's without even mentioning the Harry Potter franchise.

Above: *Snakes on a Plane* was classic 'Snakesploitation' cinema. Good for the box-office; bad for the public perception of snakes as organisms in need of our care and concern.

For many moviegoers, snake persecution like this is just a bit of harmless fun. However, snake-loving critics of Hollywood argue that these films shape societies and should shoulder at least some of the blame when it comes to the public perception of snakes in the wider world. Snake cinema may say a lot about Hollywood, but it also says a lot about us.

Above: 'I hate snakes!' – one of many movie moments when Indiana Jones faces up to his old nemesis.

Left: Harry Potter and a captive snake share a moment of emotional connection in this scene – something missing from most literary and film representations of snakes.

Snakes in the home

One final aspect of snakes in culture has become increasingly apparent in recent years – a human need for the company of reptiles that manifests itself through an ever-growing pet industry.

Estimates on pet numbers in the UK vary, but the most recent widespread statistics (2017–18) suggest that 45 per cent of the British public own a pet and the figure is likely to be rising. In all, the UK is home to 51 million pets, of which 200,000 or so are pet snakes. That figure pales into significance compared to the 11 million pet cats and the 9 million pet dogs in the UK. But it's undeniable that snakes have become a more accessible and more acceptable pet option. And pet owners who choose to keep snakes benefit from their companionship in the same way other pet owners do with rabbits, rats or any other domestic animal.

While the vast majority of snake owners are wholly responsible, not all pet owners understand the complexities of reptile requirements. As with wild snakes, pet snakes have complex habitat needs. This means that inexperienced reptile owners can end up keeping

Below: Big business: the global trade in exotic pet reptiles is worth an estimated £80 million each year.

them in substandard conditions that are detrimental to the animals' well-being, leading to disease, starvation, suffering and neglect. Not a month goes by without newspapers reporting on exotic pet snakes whose owners illegally release them into the wild or let them escape from their cage. Escapes cause undue panic to people living locally and suffering to the snake itself, which usually faces a slow death from exposure to the cold. In 2018, the RSPCA took 15,000 calls from members of the public with concerns over how to care for exotic animals, and the organisation rescued more than 500 pet snakes that had been accidentally or intentionally released into the wild.

But it's not all bad news. When properly cared for, snakes can provide their owners with a sense of purpose, companionship and warmth. And as more people choose snakes as pets and learn how to keep them responsibly, snakes gain more allies. If a little of the compassion people have for their pet snakes can be transferred to our wild snakes, which desperately need our support, ultimately that can only help with their wider conservation.

Above: In recent years, snake-handling sessions have become commonplace in schools and at birthday parties. Critics argue that activities like these send out mixed messages about snakes and the need for their conservation.

A Future for Snakes

With each passing year, we are learning how to better understand the declines in our wild snake populations in England, Scotland and Wales. All three native species are ever more isolated and are clinging on in habitats that are now a fraction of what they once were. We must do more… but what? Helping to conserve our native snakes is, in fact, easy. Anyone can become involved by recording their snake sightings, volunteering on snake surveys and by creating a snake-friendly garden.

Searching for snakes

Preserving local populations of snakes depends on knowing where they occur and recording this information. Such species records provide incredibly valuable information on distribution and are vital in helping to ensure that local habitats are protected. Recording snake sightings is perhaps the single biggest thing you can do for British snakes. Details of recording initiatives (most notably the National Amphibian and Reptile Recording Scheme, or NARRS) can be found on page 125.

Where to look

As we have seen, the best time to spot snakes is when they are basking in the very early morning in late spring and summer, or throughout the day soon after they rise from their winter slumber. A good basking site requires two things: access to the sun, and a suitable hiding place in case the snake is spooked. Many basking snakes (particularly Adders and Barred Grass Snakes) will seek out and reuse an ideal pocket of grass that is flanked on its furthest side by dense vegetation. These pockets exist naturally between grass tufts on many sites and form their own stable microclimate that is sheltered from the wind.

Above: Gardens have the potential to be a handy resource to hungry Barred Grass Snakes. A pond with plenty of hiding places can provide a superb pit stop for passing snakes, should they happen to be nearby.

Opposite: A watchful eye: future generations will judge our attempts to prevent the Adder from undergoing further local extinctions.

Above: Family-friendly reptile walks can be a useful activity for people young and old to see and better understand the important (yet elusive) animals that call their local nature reserves home.

At some sites, special strips of metal or roofing felt can be used to attract local reptiles looking for a place to bask. Later in the day, they can also serve as a handy place for snakes to hide, provided they don't get too hot. Such artificial refugia are great places to look for snakes – in fact, reptile surveyors regularly focus on them. Ant nests often form under refugia, so you may also see impressive invertebrate hunters like the Common Lizard and Slow Worm.

Surveying and monitoring

Experts are needed on the ground at many important reptile sites in order to monitor populations year on year. Ranging from trained volunteers to paid specialists, these people keep detailed records, helping to assess the fate of the snakes (and other reptile species) on a site and to raise the alarm when the beginnings of decline are spotted. They are also vital in providing crucial information on where popular hibernacula (winter refuges) may be found, thereby helping to secure the protection of these features.

Thankfully, the UK has a network of local volunteer amphibian and reptile groups on hand to help you become clued up if you are keen to get involved (see page 126). Many of these independent groups offer training events, providing you with skills that could lead to a new career in wildlife conservation.

Above: Snakes have on their side passionate and committed groups of scientists and volunteers, eager to learn more about them and help others to engage with and enjoy Britain's reptiles.

Left: The RSPB is one of many conservation groups eager to ensure the genetic mixing of populations of reptiles that exist on nature reserves.

Make a reptile paradise

Being complex in their design and multi-layered in their structure, and with plenty of potential for prey, some gardens can easily become a potential temporary hunting ground for reptiles, particularly that amphibian-hunting Barred Grass Snake. Below are some tips to help you make your garden that bit more reptile friendly.

Dig a pond

Perhaps the single biggest difference you can make to encourage Barred Grass Snakes into your garden is to dig a pond – and the bigger the better! In time, such a pond will attract frogs, toads and newts, all of which may help draw in any hungry Barred Grass Snakes that may be nearby. The RSPB website has lots of advice to get you started with your pond, including ideas on plants and potential pond designs that might suit your space (see page 126).

Below: A good wildlife pond for Barred Grass Snakes needs: plenty of hiding places; lots of sunny banks on which to bask; and plenty of amphibians – including frogs and newts.

Hiding places

Snakes don't often travel across open ground, preferring instead to move through covered areas in order to hide from cats and birds of prey. To accommodate their needs, create plenty of hiding places in your garden. Log piles are one option – the gaps between the logs are handy for snakes to squeeze into when they are trying to escape prying beaks and claws.

Above: Log piles provide lots of gaps and crannies into which snakes can squeeze themselves for protection. Some smaller snakes even squeeze into the gaps between the bark and the tree.

Below: A hungry Barred Grass Snake peers across a garden in search of a Common Frog, a particularly attractive prey item.

Rockeries

Like log piles, rockeries offer plenty of opportunities for hunting and hiding. In addition, they provide microhabitats within which humidity can build up, and they make ideal basking locations for sun-loving snakes in spring and early summer.

Above: So-called 'artificial refugia' (in this case, a sheet of corrugated metal) can attract sun-seeking reptiles. Some snakes bask on top of these 'tins' in the mornings before sheltering underneath later in the day.

Critter curtain

Few wildlife-friendly garden features can beat what I call a 'critter curtain' – a sheet of old carpet or roofing felt that can occasionally be lifted up and inspected for invertebrates. These squares of material heat up quickly, attracting colonising invertebrates like nesting ants and thereby providing feeding locations for amphibians. The amphibians, in turn, attract snakes.

Snake-friendly compost heaps

Barred Grass Snake populations are dependent on the females finding suitably warm locations in which to lay their eggs, and there are few better options than a good old-fashioned garden compost heap. If you are keen on helping local snakes out with egg incubation in your backyard space, follow the tips below.

The key thing to remember when creating a site for egg-laying is heat. In most cases, Barred Grass Snakes are interested in places that warm up nicely and that

Left: A snake-friendly compost heap needs to get lots of sun and to have accessible areas in which snakes can move in and out.

Below: When turning over your compost heap in autumn look out for the remains of the Barred Grass Snake's tiny eggshells, which resemble ping pong balls.

also retain their humidity and don't dry out too much. In terms of size, larger compost heaps tend to remain warm and humid longer than smaller ones, so bigger is always likely to be better for attracting nesting snakes. If space allows, dot several piles of cuttings around your garden to offer a choice of locations to snakes – remember that each heap will have its own unique microclimate. If conditions become very dry, consider placing a weighted-down tarpaulin (or similar) over each heap to hold in moisture.

If possible, try not to disturb a potential snake egg-laying site between June (when egg-laying may begin) and late August to September (when hatching occurs). If many eggs from different

Right: A young Barred Grass Snake: in future it may be that the public comes to revere and celebrate this garden interloper as much as it would a visiting Hedgehog or Badger.

individuals have been laid in the same compost heap (which often happens), you may observe hundreds of the tiny hatchlings hunting in your garden after they emerge in September and before they disperse into the wider neighbourhood. If all goes well, at least some of these will be back as adults, eager to find a similarly suitable egg-laying habitat to the one from which they themselves emerged.

Hibernation sites

Although many snakes return to overwinter in tried-and-tested locations, at least some may require or seek out alternative sleeping arrangements. Potentially, your garden can become one of these locations, provided you consider carefully the type and position of the hibernation sites you can offer.

Right: Artificial over-wintering sites can be bettered by having multiple entrances to encourage more reptiles (and other animals) seeking shelter in the colder months.

Over the years, reptile conservationists have toyed with various designs for hibernation sites. Generally speaking, hibernacula can be thought of as piled mounds composed of two distinct regions: an internal region with plenty of cracks, within which snakes (and lizards, frogs, toads and newts for that matter) can squeeze; and an outer layer of soil or other substrate that offers protection from frost and potential predators. Gaps between selected rocks and logs that protrude from the pile provide handy entrance and exit points to the hibernaculum.

Your hibernaculum must be in a sunny location on a well-drained site that is unlikely to flood. It should be at least 4m (13ft) long, more than 2m (6ft) wide and perhaps 1m (3ft) or more high. Ideally, the longest edge of your hibernaculum should be south-facing to allow snakes plenty of space to bask in springtime. This means they can remain close to their sleeping quarters should they become spooked by predators. Surrounding your hibernaculum with low-growing vegetation can provide cover for nearby snakes as they move in and out.

Above: 'Insect hotels' like this one generate food sources for amphibians and rodents, which in turn provide food for snakes. They can also provide excellent temporary refuges for reptiles.

Below: Reptile over-wintering sites are sometimes used by amphibians, including Common Toads (like this one) and newts.

Making connections

Snakes could once move freely and largely undisturbed across most of Britain, but this is no longer the case. As discussed earlier (see page 89), a great many snake habitats have been sliced and diced by infrastructure over the decades into a mosaic of substandard pockets. Increasingly, conservation groups are highlighting the importance of connecting existing snake populations, which is seen as the key to their long-term survival.

The RSPB is one of many organisations keen to focus

Right: Wildlife tunnels help connect bisected populations of amphibians and reptiles, provided they are maintained and not left to fill up with debris.

Below: A spectacular wildlife bridge in Banff National Park, Canada. For British snakes to persist, we may need to take more seriously forward-thinking ideas like this.

Connecting gardens

If your garden is within 100m (110yd) or so of a wetland feature such as a pond, river or ditch, then you may be lucky enough to have an encounter with a Barred Grass Snake or two. And if it is near other reptile-friendly habitats such as railway embankments, then other reptiles – including Slow Worms and Common Lizards – may also visit. To allow these animals into your garden, consider how accessible it is to visitors.

If your garden is fenced in, snakes and lizards are likely to move in and out courtesy of gaps either underneath the fencing or between the slats. If there are no gaps in your fencing, consider using a hacksaw to cut a small square out of it. Some gardeners refer to this as a Hedgehog hole, but it is equally vital as an access point to frogs, toads and newts, as well as the Barred Grass Snake, their beautiful and alluring predator.

Above: A small hole cut into the bottom of a fence is a great way to make your garden accessible to passing Barred Grass Snakes should you be lucky enough to have them in your local patch.

on species connectivity as a wildlife conservation priority. Together with the Wildlife Trusts and smaller charities such as The Amphibian and Reptile Conservation Trust and Froglife, it is making some headway. Even tiny strips of habitat – including parkland hedgerows, railway embankments and allotments – can act like bridges across urban and suburban areas. In order for snakes to thrive across Britain as they once did centuries ago, we need more of these connections. See the box above and contact the organisations listed on page 126 to see how you can help.

Celebrating snakes

Rather than being celebrated as vivid, beautiful and highly evolved mid-level predators, Britain's snakes are all too often dragged into familiar cultural clichés and stereotypes and are painted as stealthy and sly. To those who understand little about them, each and every snake is a potential threat to human life, and to those who actively despise their kind, snakes don't belong here. We'd be better off without them, these people might argue. But snakes really do belong here in Britain. They crossed the same land bridge from Europe we humans travelled over 10,000 years ago. This means that these islands are their islands as much as they are ours. And so, their great loss in numbers over the last century is a slur on our already questionable human capacity to look out for the lives of species other than our own.

But we shouldn't lose hope for Britain's snakes just yet. If I have a final message in this book, it's that all is not lost. Sure, many of the snakes that remain in populations scattered across England, Wales and Scotland are in an impoverished state, but more than at any point in human history, they are being studied, understood and included in optimistic plans to help stem the losses.

Our job as amateur and professional wildlife conservationists is to keep this turnaround going, by doing what we can in our backyards and gardens, by reporting and recording snakes, and by supporting the wildlife organisations that are applying pressure to make societal and political changes for the benefit of our wildlife.

Snakes survived the giant meteorite that destroyed the dinosaurs. With a little bit of love and care, a big heart and a loud voice of support, they might well survive us too. So, shout loud. We have many people to convince.

Above: The Smooth Snake deserves its status as a celebrity among British reptiles – an emblem of what we can achieve for Britain's rarest wildlife when we work together.

Opposite: An encounter with a wild snake is one of nature's most memorable and magical moments. If conservation efforts succeed, such moments could become ever more frequent.

Glossary

Aestivation A prolonged period of dormancy activated during hot or dry spells of weather.

Artificial refugia Specially placed materials that absorb heat, attracting reptiles for surveying purposes.

Brumation The hibernation-like state that reptiles enter during cold spells.

Cloaca The waste vent in most non-mammalian vertebrates through which excretory products and eggs or sperm are released.

Dorsal Relating to the upper side or back of an animal, plant or organ.

Hemipenes The reproductive anatomy of male snakes.

Herpetologist A scientist who studies reptiles and/or amphibians.

Hibernaculum (pl. hibernacula) A place in which a creature seeks refuge, often seasonally.

Inbreeding depression The reduced biological fitness observed in a given population as a result of closely related individuals producing offspring with one another.

Ovoviviparous Giving birth to live young. In snakes, this method of reproduction is close to, but not the same as, that of mammals.

Parasite An organism that derives benefit from another organism, often living on or within it.

Parthenogenesis A reproductive strategy involving the development of an embryo from an unfertilised egg.

Pathogen A bacterium, virus or other microorganism that causes disease.

Pellicle A thin skin, cuticle, membrane or film. In ovoviviparous snakes, offspring are enclosed in a pellicle at birth.

Pheromones A chemical signature produced and released into the environment by an animal, often for sexual signalling purposes.

Physiology A scientific branch of biology interested in living organisms and how their parts function.

Predator An animal that preys upon others.

Scale A small rigid plate that grows out of an animal's skin, often providing protection.

Serpentine movement The classic S-shaped movement that many snakes employ; also known as undulatory locomotion.

Sidewinding Method of sideways S-shaped locomotion employed by snakes that move across loose substrates such as sandy deserts.

Slough The shed skin of a snake.

Snake fungal disease (SFD) An emerging infectious disease found in numerous snake species, caused by the fungus *Ophidiomyces ophiodiicola*.

Thanatosis A state of apparent body shock, often used in 'play-dead' responses.

Thermoreceptors Specialised nerve cells that can detect changes in temperature.

Tubercle A tiny rounded projection found on the external surface of snakes.

Venom A poisonous substance intended to be injected into another animal, often for defence. In snakes, the venom is modified saliva used to incapacitate prey.

Ventral Relating to the lower surface of an animal's body.

Further Reading and Resources

Books

There is an impressive choice of books on British reptiles available to amateur herpetologists, each offering plenty of information about species behaviour, ecology and guidance on conservation. Below are a few of my favourites.

Beebee, T. 2013. *Amphibians and Reptiles*. Naturalists' Handbooks 31. Pelagic Publishing, Exeter.
In this go-to guide, Professor Trevor Beebee details the biology, ecology, conservation and identification of British amphibians and reptiles.

Beebee, T. & Griffiths, R. 2000. *Amphibians and Reptiles*. New Naturalist Library 87. HarperCollins, London.
In one of the later additions to this long-running series, respected herpetologists Trevor Beebee and Richard Griffiths collate a wealth of new and fascinating facts and observations on British reptiles and amphibians. A must-have guide for budding herpetologists.

Inns, H. 2011. *Britain's Reptiles and Amphibians*. Princeton University Press, Princeton.
This detailed illustrative guide is ideal for those eager to learn more about species identification, behaviour and, importantly, the successful conservation of British reptiles and amphibians. Useful for amateurs and professionals alike, this colourful guide is one I refer to again and again.

Minting, P. & McInerny, C. 2016. *Amphibians and Reptiles of Scotland*. Glasgow Natural History Society, Glasgow.
Produced by The Amphibian and Reptile Conservation Trust, this book offers up-to-date information about reptiles and amphibians in Scotland. It is also available as a free download from the publishers, the Glasgow Natural History Society, at www.glasgownaturalhistory.org.uk.

Speybroeck, J., Beukema, W., Bok, B., Van Der Voort, J. & Velikov, I. 2016. *Field Guide to the Amphibians and Reptiles of Britain and Europe*. Bloomsbury Wildlife, London.
British reptiles represent just a small proportion of the amazing range of species that exist across Europe. This guide includes information on ecology, distribution and key identification tips for every European reptile and amphibian species. A handy tick list at the end encourages you to get searching.

Conservation groups

The Amphibian and Reptile Conservation Trust (ARC)
arc-trust.org
ARC is an independent charity taking forward the conservation of reptiles and amphibians. As well as managing a suite of important sites for rare species, ARC has been a key player in captive breeding Smooth Snakes and successfully re-releasing them into the wild on nature reserves in the south of England. ARC also runs the Alien Encounters website (www.alienencounters.narrs.org.uk), where observations of non-native reptiles (and amphibians) can be recorded. In addition, ARC also coordinates the National Amphibian and Reptile Recording Scheme (NARRS) www.narrs.org.uk.

Amphibian and Reptile Groups of the UK (ARG UK)
arguk.org

This umbrella organisation supports local groups in many parts of the country, which themselves offer opportunities to engage in training and monitoring of reptile and amphibian populations. ARG UK also holds a national conference each year and has a website with plenty of details for local contacts and groups. It is essentially a one-stop shop for anyone who wants to engage with the local conservation of snakes, other reptiles and amphibians.

British Herpetological Society
thebhs.org

This society is one of the oldest and most prestigious of its kind in the world, founded in 1947. As well as publishing important scientific articles about reptiles and amphibians, it also supports conservation initiatives (including captive breeding) to save threatened species.

Froglife
froglife.org

Froglife is a conservation organisation that focuses on protecting amphibians and reptiles and their habitats. It runs numerous projects across the country, including creating new ponds and renovating existing ones, as well as providing practical opportunities to engage with wildlife in London, Peterborough and Glasgow.

Garden Wildlife Health
gardenwildlifehealth.org

This long-running collaborative project encourages members of the public to observe incidents of unusual mortality in garden species, including snakes. Since its inception, the project has helped scientists understand a host of complex diseases affecting garden wildlife, including snake fungal disease.

National Trust
nationaltrust.org.uk

The National Trust is dedicated to the preservation of the nation's heritage and spaces, and to the conservation of areas of natural beauty. Many of these sites are locally important lifelines for threatened reptiles, including the Smooth Snake.

Royal Society for the Protection of Birds (The RSPB)
rspb.org.uk

The RSPB works for a healthy environment rich in birds and other wildlife. As well as managing a fleet of top-class nature reserves in which threatened reptiles flourish, the organisation has also become an important player in influencing government on behalf of many of the creatures that call the UK home. Its website is a superb place to find guidance on pond building and making gardens more wildlife-friendly.

The Wildlife Trusts
wildlifetrusts.org

There are 47 local Wildlife Trusts across the UK, as well as the Isle of Man and Alderney, and together they look after 2,300 nature reserves. Most Wildlife Trusts also offer training courses, as well as running family events and local projects, including building and maintaining ponds.

Acknowledgements

First, enormous thanks must go to Julie Bailey, Alice Ward, Jim Martin and the rest of the team at Bloomsbury for making this nothing short of an utterly pleasurable writing project. Additional thanks must go to Susi Bailey and Laurence Jarvis, whose helpful comments, edits and suggestions improved the text no end.

I am profoundly grateful to a number of people who have encouraged and helped inspire my love of UK reptiles, professionally and non-professionally, either in person or through their books or guides, or simply through their kind spirit. These include Trevor Beebee, Ruth Carey, Arnie Cooke, Jim Foster, Tony Gent, Sam Goodlet, Howard Inns, Brian Laney, Tom Langton, Rob Oldham, Daniel Piec and John Wilkinson. Special thanks go to Kathy Wormald and all at Froglife for their continued support.

I came late to snakes. Although I regularly searched them out in the latter part of my childhood, I never found one. In fact, my first experience with a wild snake was probably at about the age of 23. I made up for this in my 20s, when weekend 'reptile walks' provided me with much-needed quiet time and solitude, as well as some truly memorable experiences. So attuned to snake spotting did I become in those years, that my senses still activate automatically when I hear the tell-tale crackling of tussocky grass in early spring. This appreciation of snakes came from a great reptile teacher, Roy Bradley, to whom I dedicate this book. Great teachers, professional or non-professional, change lives – here's to making more in the future.

Image Credits

Bloomsbury Publishing would like to thank the following for providing photographs and for permission to reproduce copyright material. While every effort has been made to trace and acknowledge all copyright holders, we would like to apologise for any errors or omissions and invite readers to inform us so that corrections can be made in any future editions of the book.

Key t = top; l = left; r = right; tl = top left; tcl = top centre left; tc = top centre; tcr = top centre right; tr = top right; cl = centre left; c = centre; cr = centre right; b = bottom; bl = bottom left; bcl = bottom centre left; bc = bottom centre; bcr = bottom centre right; br = bottom right

AL = Alamy; FL= FLPA; G = Getty Images; NPL = Nature Picture Library; RSPB Images = RS; SS = Shutterstock, iStock = iS

Front cover t SS, b Edwin Kats; **spine** George McCarthy/RS; **back cover** t Do Van Dijck/FL, b George McCarthy/RS; **1** SS; **3** George McCrthy/RS; **4** Mike Lane/RS; **5** SS; **6** t Raimond Spekking / CC BY-SA 4.0, bl SS, br SS; **7** t SS, b Visuals Unlimited/NPL; **8** t SS, b Alessandro Mancini/AL; **9** t George Grall/G, b SS; **10** SS; **11** t SS, b Gerard Lacz/FL; **12** SS; **13** t SS, b SS; **14** SS; **15** SS; **16** t blickwinkel/AL, b Bill Gorum/G; **17** SS; **18** SS; **19** SS; **20** t SS, b Bernard Castelein/NPL; **21** SS; **22** SS; **23** t Dorling Kindersley/G, b SS; **24** t Pete Oxford/NPL, b Hugh Lansdown/FLPA; **25** SS; **26** SS; **27** SS; **28** Warren Farnell/iS; **29** SS; **30** SS; **31** SS; **32** t SS, b blickwinkel/McPhoto/RMU/AL; **33** Carl Newman/AL; **34** Jack Perks/FLPA; **35** SS; **36** Terry Whittaker/NPL; **37** t George McCarthy/NPL, b SS; **38** SS; **39** SS; **40** SS; **41** SS; **42** Tony Phelps/NPL; **43** Adrian Weston/AL; **44** t SS, b SS; **45** SS; **46** SS; **47** Joe Blossom/AL; **48** Matthijs Hollanders/G; **49** Dom Greves/AL; **50** SS; **51** SS; **52** Kim Taylor/NPL; **53** SS; **54** SS; **55** SS; **56** Laurie Campbell/NPL; **57** Chris Mattison/NPL; **58** Geoff Scott-Simpson/NPL; **59** Guy Rogers/RS; **60** James Lowen; **61** Roger Tidman/FLPA; **62** SS; **63** SS; **64** SS; **65** Jdira Simo/G, b Paul Sawer/FLPA; **66** SS; **67** Sergey Uryadnikov/AL; **68** t © Biosphoto, Christian Gautier/FLPA, b David Perpinan/NPL; **69** Pollwog/AL; **70** Jack Perks/RS; **71** SS; **72** Andy Sands/NPL; **73** Andrew Chamberlain/AL; **74** Sven Zacek/NPL; **75** cwk15/iS; **76** t SS, b Jules Howard; **77** t Tony Phelps/NPL, b SS; **78** t viktor2013/iS, b SS; **79** Tony Phelps/NPL; **80** Tony Phelps/NPL; **81** t Daniel Heuclin/NPL, b mgs/G; **82** Carl Newman/AL; **83** t Daniel Heuclin/NPL; **84** David Kjaer/NPL; **85** John Cancalosi; **86** t Mark Waugh/AL, b Lillian Tveit/AL; **87** SS; **88** Richard Revels/RS; **89** t SS, b Michael Schellinger; **90** t wildfocus stock/AL, b Patrick Fox/AL; **91** t Kieran Grasby/G, b SS; **92** Les Gibbon/AL; **93** David A Eastley/AL; **94** SS; **95** Cyril Ruoso/NPL; **96** Nina Vetrova/AL; **97** DU Photography/AL; **98** t artincamera/AL , b SS; **99** jeffowenphotos/'Snake' by Sidney Nolan/Creative Commons; **100** Michelangelo/Public Domain; **101** Acquired by Henry Walters/Public Domain; **102** INTERFOTO/AL; **103** Jim Champion/Brusher Mills' gravestone/CC BY-SA 2.0; **104** Library of Congress/Public Domain; **105** t Clark Stanley's Snake Oil Linment/Public Domain, b Snake Soup/CC BY-SA 3.0; **106** Unitwed Archives GmbH/AL; **107** t AF arcive/AL, b Collection Christophel/AL; **108** Jack Taylor/AL; **109** SolStock/iS; **110** Tom Mason/RS; **111** Joe Blossom/AL; **112** Nick Upton/RS; **113** Tom Mason/RS; **114** SS; **115** t SS, b Joe Blossom/AL; **116** David Tipling/NPL; **117** t SS, b Andrew Darrington/AL; **118** t Sandra Standbridge/RS, b Roger Parkes/AL; **119** t David Broadbent/RS, b SS; **120** t SS, b Steve_Gadomski/iS; **121** David Hosking/FL; **122** James Lowen; **123** Paul Sawer/RS.

Index